COOLHAM AIRFIELD REMEMBERED

Memories and anecdotes of
a Sussex D-Day Fighter Station and Village

Paul Hamlin Paul Hamlin

and

Ann Davies

Cover design by Paul Hamlin

Typeset by West 4 Printers Ltd,
8 Essex Place,
London, W4 5UT.

Printed by Woolnough Bookbinding Ltd,
Express Works,
Church Street,
Irthlingborough,
Northants, NN9 5SA.

Published by Paul Hamlin,
Fairlands Cottage,
Two Mile Ash,
Horsham,
Sussex, RH13 7LA,
England.

ISBN 0 9527968 0 5

Front cover photograph: Crew of B-26 Marauder
"Flounder Gus".

This book is dedicated to all who served at
Coolham Advance Landing Ground Airfield
in particular the following
who courageously gave their lives –

Roman Marian BRYGIDER
Stanislaw CALIŃSKI
Bolestaw GECA
Czeslaw OBERDAK
Kazimierz OSIELENIEC
Henryk STEFANKIEWICZ
Adam SWORNIOWSKI
Wladysław Jan SZAJDA

Charles GORDON
Noel Sturt GREEN
Allan John HANCOCK
Douglas Haig NELSON
Stanley PAYNE
William Edward RIGBY
William TYRER

No.133 and 135 Wings, 84 Group Second Tactical Air Force

POLISH AIR FORCE
306 Dywizjon Myśliwski, „Toruński"
(306 Fighter Squadron)
315 Dywizjon Myśliswki, „Debliński"
(315 Fighter Squadron)
411 Eskadra Reperacyjna
(411 Repair and Salvage Unit)
5090c Ruchoma Jednostka Lacznośći
(5090c Mobile Signals Unit)

ROYAL AIR FORCE
Fighter Squadrons
129 Mysore, 222 Natal, 349 Belgium, 485 New Zealand

Airfield Construction Squadrons
5003, 5004, 5018, 5027, 5206

RAF Regiment
2722, 2829 Squadrons

RAF Ordnance Unit
422 Aviation Fuel and Ammunition Park

RAF Maintenance Unit
49 MU (Faygate)

Photo: RAF Museum Ref No. P021018

Foreword

by

Wing Commander David Cox
Commanding Officer, 222 (Natal) RAF Fighter Squadron
on behalf of British, Allied and Commonwealth airmen

Coolham airfield is not a famous name such as Biggin Hill or Tangmere of Battle of Britain fame. However, from April to July 1944 it was "home", although rather primitive, to several hundred men of the Royal Air Force in operations against the German fortress of Europe prior to the D-Day landings and afterwards.

Six fighter squadrons operated from Coolham, three Mustang and three Spitfire. Their pilots and ground crew came from many nations – Poland, Britain, Canada, New Zealand, Australia and Belgium.

There were casualties, especially amongst the Mustang squadrons who suffered terrible losses.

Coolham airfield was an operational base which then reverted back to its original farmland.

Its rebirth was started by Paul Hamlin who, as an Air Training Corps cadet had heard talk about a wartime airfield south of Horsham.

Later, he started a four year research into the history of this mystery airfield. The result was, 50 years on from D-Day, a wonderful two-day reunion in June 1994 of many of those who had served at Coolham, the dedication of a most moving memorial near the local pub and, a year later, an historical location sign at the entrance to the old airfield.

This book is about Paul Hamlin's research into the history of Coolham airfield, plus accounts from some of those who served there.

Why did Paul do all this? The answer is simple. He wished to honour all those, both in the air and on the ground, who contributed to a victory which ensured that for his and future generations there would be freedom to enjoy the Sussex countryside.

Wing Commander D.G.S.R. Cox, DFC and Bar, C de G,
Berkshire, England.

Photo: M. Cwynar

Foreword

by
General Stanislaw Skalski, Officer Commanding
133 Wing, 84 Group Second Tactical Air Force,
on behalf of Polish airmen

On the fiftieth anniversary of the D-Day Normandy Landings, a poignant ceremony took place in the centre of a peaceful, West Sussex village.

As a tribute to Allied airmen, who lost their young lives while flying from the nearby RAF Landing Ground of Coolham, a magnificent war memorial was unveiled.

I was cordially invited to be there.

That memorable event was a culmination of one man's effort and determination to preserve for posterity the memories of wartime drama that had taken place from the nearby airfield.

He painstakingly researched about events of 1944 forming a committee to organise fundraising towards this worthy memorial with the help of a notable artist, Cedric de la Nougerede, who painted Dziubek Horbaczewski's Mustang and sold hundreds of prints for the memorial fund.

Substantial residual funds were later donated to English and Polish charities.

With a feast of dates, and knowledge accumulated during years of research, Paul, together with his contributors, with easy style, and nice touch of phrase, vividly recreates Polish and Allied airmen's interaction with the patient local community, describes the drama that had taken place from rural Coolham so many years ago and preserves for future generations and posterity, historical events of those bygone momentous years.

General Stanislaw Skalski, DFC, DSO, VM,
Warsaw, Poland.

Coolham Airfield Remembered

CONTENTS

Introduction

To think of Coolham, a tiny village in the heart of West Sussex, as host to an operational RAF station, with fighter planes flying in and out, stretches the imagination.

But in wartime many strange things happen and Coolham found itself right in the front line of defence as one of several Advance Landing Grounds (ALG's) used in the run up to D-Day, 6 June 1944, and for a few weeks afterwards. They were temporary airfields built to provide extra air support during the invasion of Europe, code named Operation Overlord.

Gathered together in this book are the recollections of airmen who served at Coolham ALG during that momentous time in history and also the memories of civilians, villagers and those who were children at the time and regarded it all as very exciting.

And what child wouldn't have been captivated at the site of an airfield being constructed right under their noses, with dashing pilots from multi national squadrons roaring off in Spitfire and P-51 Mustang fighter aircraft.

Some of these pilots were already heavily decorated heroes of the Battle of Britain including one who had also been awarded the Order of Lenin, the highest honour the Russians could confer, for his service when sent out to their country.

This was history in the making.

Many of the anecdotes recalled are amusing, others downright funny, highlighting the camaraderie and indomitable spirit of service personnel and civilians at this dangerous time.

Village boys palled up with ground crews and chalked on bombs slogans like "Present for Adolf" and "Goering, this one is for you".

The pilots have improbable tales to tell such as Squadron Leader Horbaczewski's heroic act of rescuing a fellow pilot from his crashed aircraft in a marsh near Cherbourg and returning to Coolham with him in his single seater P-51 Mustang.

And one hilarious anecdote is from a Polish pilot who appeared to land his Mustang backwards! But more of that later.

Two American bombers made emergency landings there and one crew member recalls going to a dance in a nearby village, having only his flight kit to wear, trying to be nimble footed in his flying boots.

It obviously wasn't all fun and games and the serious, tragic side of the war took its toll among the gallant young pilots and other airmen at Coolham, 15 of whom were killed.

It was when Paul Hamlin discovered this that he was spurred on to organise a fitting memorial to them so their names and sacrifice would never be forgotten.

And this memorial now stands outside the Selsey Arms public house in Coolham, where pilots and groundcrews used to gather for a relaxing, off duty drink . . . or two.

A.D.

1

The Project's Beginnings – Paul Hamlin

In 1968, as a 14 year old member of 1015 Squadron (Horsham) Air Training Corps, under the leadership of F/Lt Gilbert Parr, I learnt that there had been an RAF Fighter Station in the Coolham area during World War II but nowhere was there any information relating to it.

Having joined Sussex Police in 1975, being posted to Horsham in 1980, it was not long before I felt the urge to find the airfield. On examining an old map at the Police Station there appeared to be lengths of perimeter track marked as a public bridleway leading from the B2139 Thakeham Road, approximately 200 yards south of the A272 Coolham Crossroads.

On a quiet sunny morning in the summer of 1981, my duties took me to Coolham. After a short walk along the public bridleway, the sight of acres of wheat sweeping into the distance greeted me, growing on what had been the site of the main 1,500 yard E/W runway – the exact spot where an historic event had occurred, where men from distant lands had been called upon to act with extraordinary courage and endurance while fighting to preserve freedom, providing air support prior to, and during, Operation Overlord, D-Day, 6 June 1944, the largest and most complex military invasion in the history of mankind.

It was difficult to imagine that the now tranquil countryside beneath the beautiful South Downs had once been the site of a fully operational RAF Station where young men lived in cramped, basic conditions under canvas, flying powerful fighter aircraft on perilous missions into occupied Europe.

Yet there were no official signs to mark the location of the airfield or a memorial dedicated to the personnel who had served there.

Something told me that the situation had to be redressed. I was convinced that there must have been some significant happenings during this brief period of activity and was determined to try and find out exactly what had occurred.

Enquiries were made at Horsham Museum, but to no avail, nor were there any reference books detailing the squadrons which operated from Coolham or similar temporary airfields. Its existence was confirmed in 1988 as a result of purchasing a copy of *Action Stations 9* by Chris Ashworth, Patrick Stephens Limited.

With my keen interest in the airfield I was always eager to attend calls in the Coolham area, hoping to obtain more information. On Saturday, 22 December 1990, I met Keith Leech, proprietor of J.K. Leech Fuels, Coolham, where a burglary had taken place. He introduced me to David Cleverley, manager of the Coolham Crossroads Garage, an ex-member of No. 2 Group Royal Observer Corps, Horsham.

As a police officer I was in the fortunate position of meeting residents who had knowledge of the airfield, a number of whom had wartime memories. Those who were there during hostilities assured me that the countryside's character and appearance had hardly changed, the roar of the Merlin engines being replaced by skylarks and occasional aircraft using Gatwick.

It soon became clear that in order to satisfy my curiosity as to what had officially occurred, a detailed investigation would have to take place. This led to numerous visits to the Public Records Office, Kew, to obtain photocopies of each Squadron's Operations Record Books, studying documents at the RAF Museum, Hendon, and corresponding with Squadron Associations, Government departments, museums and individuals worldwide.

In response to an article in the *West Sussex County Times,* dated 14 February 1992, by reporter Jan Hatwell, entitled *Secret Strip's Vital Role,* based on limited information in my possession, an excellent response was received from readers who either knew of the airfield or were interested in its history.

One person who wrote to me was Michael Tanner-Garbarz, an ex-Coolham airman from Pulborough. I am adamant that without his assistance, and also of Kazimierz Barczak, representative of the Polish Air Force Association, London, many contacts would have remained untraced.

With correspondence flooding in the project gradually took over my life. Research revealed the airmen had to overcome inherent language, cultural and logistical problems while converting to new aircraft in an extremely short period.

An efficient fighting unit quickly developed resulting in extraordinary stories of endurance and heroism. It seemed a terrible oversight that so much had happened on our doorstep and yet there was nothing to indicate that these events had ever occurred; as if the airfield and its airmen had never existed.

I was astonished to discover that 174 pilots were stationed at Coolham ALG and, of these, ten fought in the Battle of Britain. While serving there 15 personnel died in the course of their duties, eight Polish, six British and one Canadian.

In May 1992, a committee was formed to raise funds and to organise a Coolham Airfield Remembered Community Commemorative Event on 11/12 June 1994, the 50th anniversary of its use.

During the historic weekend a Memorial was unveiled in the garden of the Selsey Arms public house, Coolham Crossroads, in the presence of returning veterans. The project was registered with the Organisation and Planning Department of the Sussex Police, who gave their support to my private venture.

One of the most gratifying things about researching and organising the Coolham Airfield Remembered Event has been the friends and acquaintances I have met world wide who have welcomed me into their homes.

Without doubt my experience of a low level flight over beautiful British Columbia in a Cessna piloted by Mike Humphries of 129 (Mysore) Squadron weeks before his tragic death in a flying accident is unforgettable.

It is pleasing that so many friendships forged by conflict have since been renewed in peace, the true spirit and camaraderie between veterans is still evident.

I would like to take the opportunity to thank them for their contributions towards this book by providing reference material in the form of treasured personal memorabilia, including pilots' log books, letters and photographs, which have been invaluable.

This publication would not have materialised without the professional assistance of Ann Davies and John Sampson, chief photographer of the *West Sussex County Times* who meticulously enhanced the ageing prints.

It was always my intention that the human interest testimonies related to me should be published in a non technical, anecdotal format, detailing the day to day experiences of those who served at Coolham ALG together with local residents whose quiet rural lifestyle was disrupted.

On no account did I wish to glorify war. Now my vision has been realised I sincerely hope future generations will look at the Memorial and view the 15 commemorative oak trees beside the public bridleway on the northern section of the ALG, remembering the gallantry and sacrifices made in the name of freedom.

Never again will Coolham be the forgotten airfield.

Paul Hamlin at the Coolham Airfield Memorial *Photo: J. Sampson*

2

Constructing the Airfield

RAF Coolham was situated six miles southwest of Horsham, one of seven Advanced Landing Ground airfields (ALGs) constructed in Sussex for the use of Fighter Command and its allies. The others were at Appledram, Bognor, Chailey, Deanland, Funtington and Selsey.

Its primary purpose was to provide air support to forces liberating Europe prior to, and during, Operation Overlord: D-Day, 6 June 1944. The airfield was situated within 11 Group District with its headquarters at RAF Uxbridge.

During June and July 1942, Air Ministry surveyors scoured southern England looking for suitable sites for ALGs. A total of 72 were located, however, as a result of fierce opposition from the War Agricultural Executive Committee, this figure was considerably reduced to 23 actually being built.

On Thursday, 8 October 1942, a top level Air Ministry conference was held at Coolham when it was finally agreed an airfield would be constructed on land to the east of the B2139 Thakeham Road with Saucelands Lane adjoining its southern perimeter.

By virtue of its level terrain and the fact that the area had excellent road and rail links, namely Billingshurst, Christ's Hospital, Southwater and Horsham, it was an ideal location.

A total of 113 acres of agricultural and wooded land was requisitioned. It was estimated that between 700–800 hedgerow trees would require clearing and 2,800 yards of HT electrical cable be buried, together with wiring for telex and telephone facilities (PBX).

It was initially thought that the preparation of the site would be slow as under-drainage was required in view of the flatness of the strip area and heavy subsoil.

During August 1943, 156 personnel of Nos. 4661 and 4663 AC Flight, 5004 Airfield Construction Squadron, commanded by S/Ldr J.K. Lancaster, commenced construction of the airfield, a task that took 10 months. Also present were a small contingent from 5206 (Plant) Airfield Construction Squadron.

According to 5004 Squadron records the following plant was used: two David Brown Tractors, three Dumpers, one DH10 Bulldozer, one D8 Bulldozer, two D4 Bulldozers, one Water Trailer, one Concrete Mixer, one portable saw, one Petrol Bowser, one Scraper 12 yard, one Blade Grader, one Roller 8 ton, one Roller 2½ ton, one D4 Tractor, one Disc Harrow, one Tractor Harrow, one Gang Roller, one Agricultural Trailer, one Paint Sprayer.

Barn House Farm, Shipley, situated at the southern end of the airfield, was a POW camp controlled by the Army. A number of airman constructing the ALG were billeted there.

Local contractors H. Kay, of Horsham, were also employed in initial ground clearance and levelling. Ronald Coleman, a civil engineer seconded to the War Department, worked with RAF and civilian personnel at the ALG from its inception.

West Sussex County Council objected to the demolition of Sauclands, an historic building which was subsequently saved.

However, an adjacent property, Five Mile Ash, situated on the south side of Saucelands Lane, was demolished due to it directly obstructing the southern end of the N/E–S/W runway. Bridge Hill House, Thakeham Road, Coolham, was positioned close to the western end of the main runway and for safety reasons its roof was removed making it uninhabitable and was restored after hostilities.

Two Sommerfeld metal strip runways were laid along with 36 aircraft dispersal hard-standings, the main E/W runway being 1,500 yards in length and the S/W–N/E runway 1,200 yards in length.

Barriers were situated in Saucelands Lane closing it to the public when the S/W-N/E runway was in use.

The control tower was a moveable caravan and the ALG was assigned the airfield identification letters XQ. There were no night time landing aids, sorties were monitored by personnel from No. 2 Group, Royal Observer Corps, Denne Road, Horsham.

Four Blister aircraft hangars were erected along with pyrotechnic dumps, an armoury, defence post and a petrol installation for 24,000 gallons. RAF Fire and Rescue Services were also present. Tents and marquees were used as squadron offices, guardroom, stores, motor transport buildings and a canteen. Medical facilities were provided in a mobile surgery and tents. Part of Farleys, situated on the airfield, adjacent to the Thakeham Road, was used as a nursing ward and administration centre.

Virtually all airmen who were stationed at the ALG and were operational lived in tented accommodation on the airfield. Conditions were sometimes cramped, cold and wet with a number of NCO airmen having to use wooden boxes on which to sleep. Bathing and toilet facilities were also extremely basic.

As the majority of the airmen did not possess motorised transport, during their brief off duty periods they normally relaxed by visiting the nearby Selsey Arms public house at Coolham Crossroads or cycled to nearby villages.

Paul Hamlin

❑❑❑❑

One of those involved in constructing Coolham ALG and laying the Sommerfeld track for the runway, was Leading Aircraftman Bill Winchester of 5004 Airfield Construction Squadron.

At Coolham we lived in a row of houses close to the site, possibly council property taken over by the Air Ministry with the land.

We generally laid out the sites, levelled the ground, felled trees to the approaches to the runways and laid the Sommerfeld. The track was like large mesh chicken wire, in rolls 10 or 12 feet wide with round iron rods running across its width and looped at each end.

When rolled out, flat iron bars were threaded through the loops and all secured to the ground by two and three foot long iron pickets which were driven into the ground by three men wielding sledgehammers in perfect harmony.

When one length of the runway had been secured other rolls of track were added for the required width of the runway.

A tractor was then used to stretch the width of track while it was then secured until the whole length was taut.

Grass seed was then sown over the area and there was your runway.

It was certainly a hard day's work but at 22 years old it was much better than being on a station and very healthy too.

It was a great pity that we never saw the end result of one of these airfields that we began. We never saw one of them operational.

At Coolham I remember the large house by the lake and the invitations to Harsfold Manor, Wisborough Green, the evacuated offices of an insurance company.

We also put on a dance at Billingshurst.

When one looks back they were good days really.

Bill Winchester, LAC 1289132
Eastbourne

Another of those involved was Corporal Fred Hill, of 5206 (Plant) Airfield Construction Squadron.

I arrived at Coolham with my unit of some 50 men from 5206 Squadron in February 1944, to find our living quarters were to be in farm buildings in Saucelands Lane.

As a temporary measure, and as I lived at Wisborough Green, I applied to the Commanding Officer for a living out pass and cycled to and from home each day.

The next day we got straight down to laying the Sommerfeld track which we were highly trained to do.

There were hundreds of rolls of track to be moved to the sites and being the only plant operator it was my job to drive the crane. The rolls of track were much too heavy to be manhandled – it would take six men to move each roll into position.

5004 Squadron were already there and had already laid a good part of the runways and buildings were springing up like mushrooms.

We were given the job of completing the end of the runway at Saucelands Lane and a number of aircraft dispersal sites which ran alongside the Thakeham Road and the perimeter track by Farleys Farm.

I remember the American Marauder coming in which we thought was a dummy run but turned out to be a crash landing onto this newly laid runway causing considerable damage, ripping 100 yards of track up and tragically killing one of the personnel of 5018 Squadron.

I also remember the house at the Coolham village end of the runway being demolished to clear the flightpath for landing and take off.

Coolham airfield did not come into operation by the time my unit was suddenly posted to Scotland on 28 February 1944.

Fred Hill, Cpl 1726865
Bogner Regis

Cpl Fred Hill

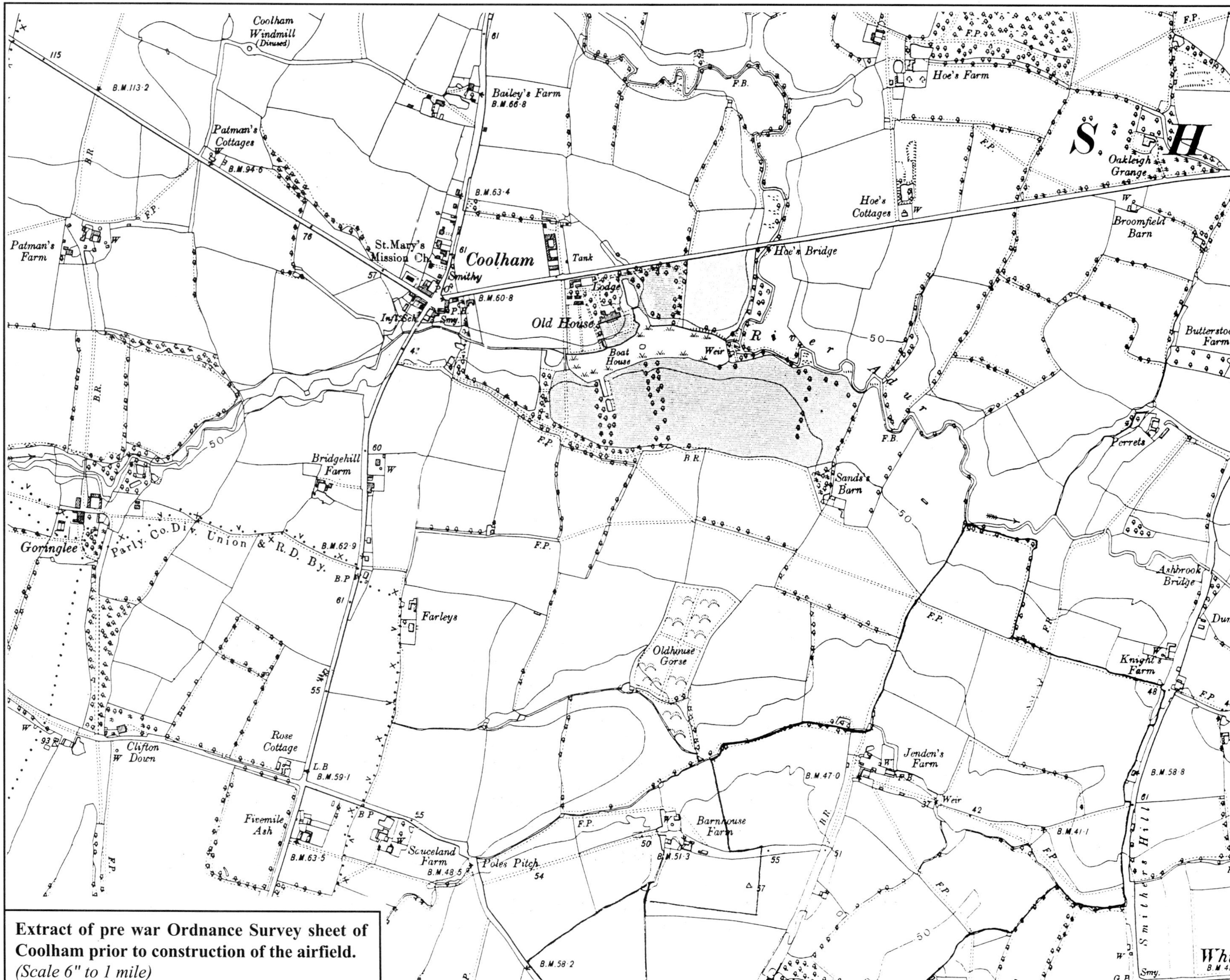

Extract of pre war Ordnance Survey sheet of Coolham prior to construction of the airfield.
(Scale 6" to 1 mile)

Crown Copyright RCHME

Aerial view of Coolham ALG, 1 January 1944

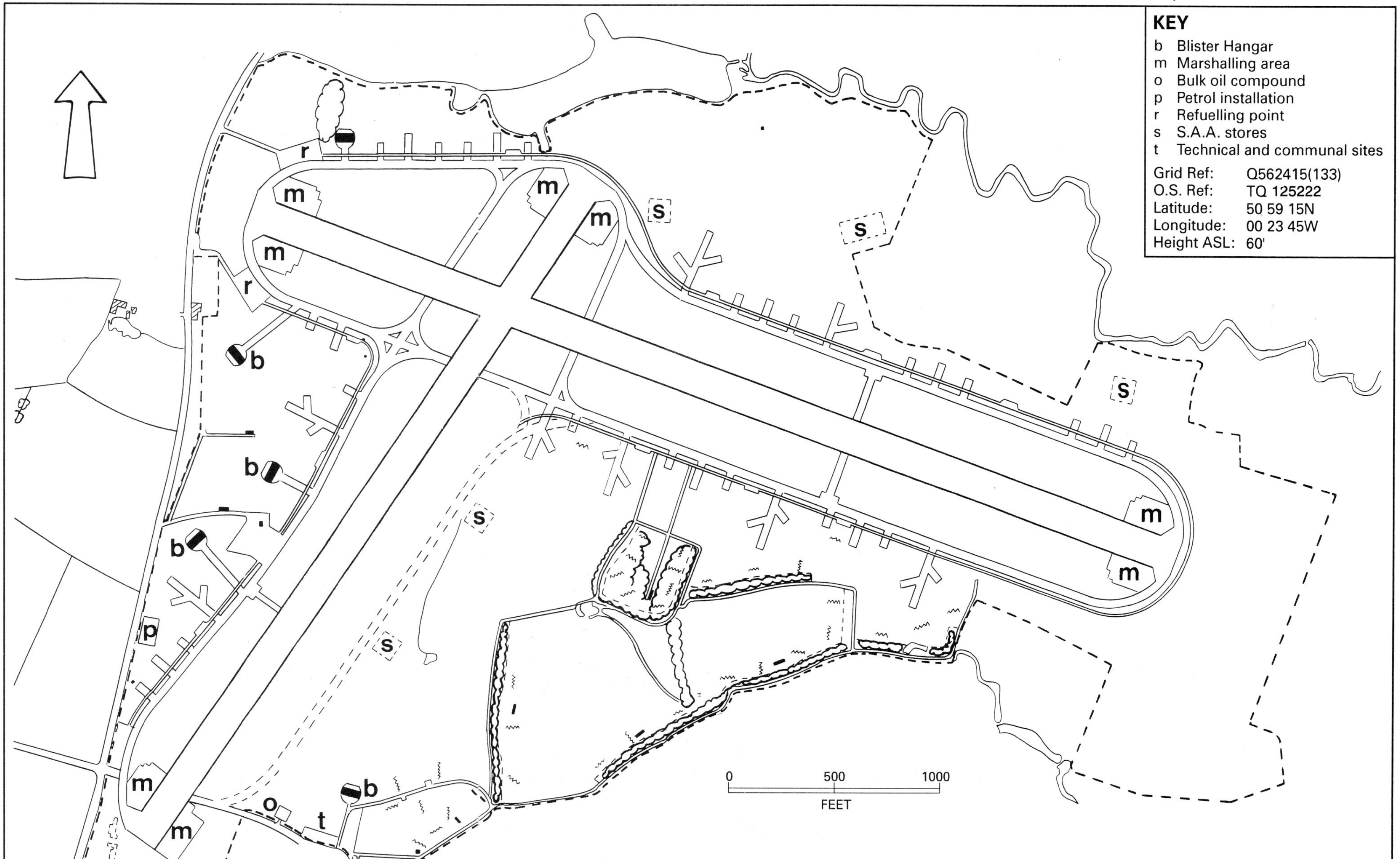

COOLHAM ADVANCED LANDING GROUND

Drawing derived from Air Ministry Plan C-71

Crown/RAF Museum

P. Francis

Photo: K. Barczak

3

General Skalski – A Strange Flight

There is no finer account of the tension felt by pilots on their dangerous missions from Coolham ALG than that of General Stanislaw Skalski, DSO, DFC, VM, who was Wing Commander of 133 Wing, 84 Group, Second Tactical Air Force.

He held the distinction of being Poland's top scoring fighter pilot of World War II having 21 confirmed victories.

Skalski's record of one particularly hazardous flight on 19 May 1944, Ramrod Mission 894, puts perfectly into perspective the nerve tangling, knife edge judgment they had to exercise all the time while in the air and the irritation and exasperation when they felt things were not quite right.

If anyone is in any doubt about the crucial role played by Coolham ALG and its airmen this should set the record straight.

They might have larked around on the ground and made light of their escapades but in the air it was a different story where sheer instinct too had a hand in survival. He writes:

An enormous High Pressure, which ranged from the shores of the Atlantic Ocean to Poland, resulted in very favourable weather conditions to begin a gigantic air offensive against the German Reich.

May of 1944 was the last "quiet" month before a decisive strike against the Atlantic Wall was made. Thousands of Allied bombers taking part in the night and day operations softened the ground before the major onslaught.

Two Polish Fighter Wings (131 and 133) were ready for immediate action.

I commanded the 133 Wing, which was based on an Advance Landing Ground at Coolham, near Horsham, 50 miles north of Southampton. Every day we took part in long range flights, escorting American Flying Fortresses and Liberators on their missions to attack strategic targets deep in the heart of the Third Reich.

It was a memorable day in May when, from the early morning hours we were receiving a number of conflicting orders from our Headquarters. Something was not quite right. Our ground crews were working hard alternatively mounting long range tanks or the 500lb bombs.

To make things worse, after two weeks of typical hot African weather, the sky was invaded by dark, low clouds. The changing weather was one reason why conflicting orders were being issued.

Finally, at 8am, a messenger brought us the final order, which was the "D" form. Although in appearance it resembled a roll of toilet paper the "D" form contained secret information pertinent to the ensuing operation.

I set the briefing time for 8.30am in the Operations Room, which in reality was a large tent suspended between two heavy lorries. Pilots from the 306, 315 and 129 squadrons filled the room.

I looked at them from a raised platform and saw a group of young, suntanned, determined faces, whose excited eyes, directed towards me, were anxious to know where we were going and what was our target.

My briefing was naturally conducted in English, because our Wing consisted of two Polish and one British squadron and because in the air we had to communicate in English.

Addressing the squadrons I told them –

"Our objective is as follows: We are assigned to escort the middle box of the Fortress formation on their way back from Hamburg – our target today.

"We will rendezvous with the Fortresses over Hamburg at 12.00 hours. When over enemy territory I will order the Wing to form the battle formation, i.e. each squadron with three sections of four aircraft, and on meeting with the Fortresses we will position ourselves on the right side of the box.

"Strict radio silence, until over enemy territory, shall be observed. I will lead the 306 Squadron, close escort, at an altitude of 30,000 feet. Above us will be 129 Squadron as middle cover and above them, 315 Squadron as top cover.

From Coolham to Hamburg our route will be Coolham–Dover–The Hague–Hamburg.

"All section leaders will receive maps with calculated courses and times. From The Hague, at an altitude of 30,000 feet, we will face quite a strong headwind.

"At present the cloud ceiling is 600 feet with layers of clouds ending at 24,000 feet. We will climb, each squadron in tight formation, until we are through the clouds, at the speed of 220mph.

"We will enter the cloud layer simultaneously, line abreast, with each squadron one mile apart from the other.

"When we break through the clouds over the English Channel we will set course for The Hague. Take off time 10.10 hours. Please synchronize your watches.

"We will take off in the following order: 315 – the top cover squadron, 129 – the middle cover, and 306 – the lead squadron. Any questions?"

There were no questions, only silence.

Everybody, a little nervous, began talking at once while collecting maps and route information from the Intelligence Officers and then proceeded by car or on foot to their respective dispersal areas.

All around the airfield engines could be heard, either still idling quietly or emitting a mighty roar of the maximum rpm. Soon the noise became unbearable and the nearby wood reflected their echo into the surrounding countryside.

The Mustangs, with the underslung long range tanks, began rolling towards the starting location on the steel-net runway.

Each squadron was on the runway in take off position before the signal was given.

Suddenly, a green rocket was fired and the first squadron took off, then the second and finally, 306 Squadron.

Before I completed the first turn over the airfield, I had the whole formation in place.

As we approached the clouds I set the course east and saw all the aircraft flying in their respective formations, each of them in their proper place. We entered the clouds and it was quite dark. We continued to climb, still wrapped in by a thick, never ending, cloud. The altimeter indicator was moving steadily, showing our progress upward.

From the corners of my eyes I could see my wingmen.

We were experiencing a little turbulence but the heavily loaded Mustangs took it in their stride. The flight was quite tranquil – there were no surprises.

As we gained altitude there was more light and the surroundings seemed more pleasant; pilots liked that much better and they felt much happier.

Suddenly the Mustang's body shuddered. Not to worry, we just went through 21,000 feet, and the second supercharger kicked in, increasing the engine's power. In our ears we could feel the difference – the pleasant, more rhythmic sound of the classic engine.

It was getting very light and sunrays began to penetrate the clouds.

Outside of the cockpit was quite wet. Radio silence continued to be strictly maintained.

I realised that all pilots were under tremendous stress but nobody interrupted the radio silence.

On everybody's mind was the thought that somewhere close by there were two other squadrons trying to break through the cloud cover at the same time.

Any deviation from the set course could end in a disaster – air collisions have been known to happen during a flight on instruments.

Nothing was heard from the Ground Control, nothing interrupted the deadly silence, except the regular hum of the engine.

We had reached 24,000 feet. It was very light and yet we were not through the clouds.

I knew that everybody was very impatient, for we had reached the estimated altitude.

However, our pilots were very skilled in formation flying and therefore would not dream of breaking formation while still in clouds.

Suddenly, at 25,000 feet, I seemed to have jumped out from the cloud, as though from a catapult, and immediately began to look around for the rest of the formation.

Like black beetles, Mustangs were popping out from the clouds below. The two other squadrons, still in parade formation, appeared on my left and my right, respectively.

I checked my watch – we were on time. I set the course for The Hague holding a map in my lap, knowing that it would come in useful, but then I realised that the cloud cover extended all the way to the target and the dead-reckoning navigation was out of the question.

I put the map away and concentrated only on holding the calculated course setting and on the time checkpoints.

We were flying at an altitude of 28,000 feet. I would have preferred to fly higher because, from experience, I had a lot of respect for the German ack-ack. Besides, higher would be the best altitude for our Mustangs.

In a flight over the unending cloud cover the rythmic sound of the engine and the monotonous panorama created boredom. There were no landmarks for one's eye to observe, nothing but an expansion of a great white ocean below and the blue sky above.

Time passed so slowly, although our true airspeed was steady. Every few seconds I glanced at my watch to check whether it was time to change the course. We were still a long way from the target.

Four minutes more and I set course for Hamburg, at 109 degrees.

Since nothing was happening around us I chose not to announce changing of the course on the radio, but simply made a gentle turn to the right and proceeded along the new course. The sun was now above and a little to our right.

The middle cover (129) and the top cover (315) squadrons, after changing over their respective formations, took their new positions on my left, in two echelons, all the way to 30,000 feet.

As soon as we had completed our manoeuvre we were greeted by the German artillery fire (88mm cannons), black smoke puffs exploding below at about 25–26,000 feet. Soon the salvos became more frequent, more accurate, approaching our formation at the correct altitude and perhaps a little above and in front of us. I began gentle manoeuvres changing our speeds and altitudes.

"More gas everybody," I announced over the radio. After a few isolated salvos behind us we were out of their range.

In the sky above us there was no sign of the German Luftwaffe. Our Ground Control was completely silent. By this time we might be already out of their range.

Steadily we were approaching German territory, not far from the Ruhr region, which meant concentrated anti-aircraft defence.

I surveyed the sky for signs of the German fighters. I was positive that they would not let us through that easy.

Suddenly, radio silence was broken. "Hello Oxo Leader, Carmen one eight – a formation of bogey (unidentified) aircraft at 3 o'clock below!"

I looked over and indeed I saw over 30 black spots rapidly gaining altitude.

"Oxo Leader speaking. They are probably ours since they are flying on the same course as we are. Keep them under observation."

The bogey aircraft did not try to close their distance between us but continued flying on their own course.

They were probably ours, flying towards the target with possibly the same assignment as ours. For some time we were flying in parallel.

Suddenly I saw black smoke puffs as several accurate artillery salvos were being fired at the bogey aircraft below us. At once the whole formation scattered all over the sky.

I could not tell whether any of their aircraft had been hit though after a few seconds I could no longer see any aircraft below us at all.

In order to avoid the German ack-ack I made a slight turn to the left, changing the course by four degrees, from 109 to 105 degrees.

After a few minutes there was a concerned reaction on the radio waves. It was Dziubek (S/Ldr Horbaczewski), Squadron Commander of the 315 Squadron, enquiring: "Hello Oxo Leader, Rainbow 14 speaking. Is your compass OK? What is your course?"

"Hello Rainbow 14, Oxo Leader speaking. Course 105 degrees, my compass is OK!"

But Dziubek was not convinced.

"Hello Oxo Leader. Why are you changing course?"

"Hello Rainbow 14. Don't Worry!"

Actually, Dziubek was right to worry and to question my decision because, if we did not keep the original course, which was calculated taking into consideration the speed and the wind direction, where the heck would we end?

And I had absolutely no reason to change the course in the first place, especially that without seeing the ground there was no way to confirm our position, nor did I have any additional information from our Operations Room, which continued to remain silent.

At 11.30 hours I changed the course again by five degrees to the bearing of 100 degrees. This time my decision was too much for the Squadron Commander of the British 129 Squadron, S/Ldr Haw.

"Hello Oxo Leader – Bad Hat 14 speaking. What is happening? Why have you changed the bearing again? We will miss our target!"

With a noncommittal reply I managed to pacify Haw and although it was quiet again I could only imagine how worried must everybody be, listening to this conversation and thinking thus: "We were still half an hour away from the target and our Wing Commander is leading us into an unknown territory. With a considerable distance away from Hamburg, any small deviation from the course could have tragic consequences. Where the heck is he going?"

As for myself, I was convinced that the new bearing selected by me was the correct one. So I thought, "Let them worry a bit, they still have to follow me! There must be strict discipline in the air! Without teamwork we cannot expect positive results! After all, I am responsible for executing the task assigned to me by my Command."

In spite of my defiant thoughts, I was still sweating it out, thinking that perhaps they had been right and that I might have been wrong. After all, I was responsible for the 36 aircraft and the 36 human beings – my experienced and brave pilots.

I agonized with my negative thoughts, but I stuck to my last bearing. There was not much time left, for we were only 10 minutes away from the designated target.

What if I did make a mistake? What if we were already over the North Sea flying towards Denmark instead? I did not want to believe that.

At 11.54 hours – there were only six minutes left. I had to control my nervous tension! After a while, however, on the horizon ahead, I spotted some aircraft sparkling in the blue sky, but then they disappeared into the cloud cover.

With confidence I did not feel I decided to break the heavy silence.

"Oxo Leader speaking – everybody look out – observe the horizon. The target is in front of us. In three minutes we will be at the rendezvous region."

In the meantime, I had been straining my eyes, scanning the sky for any sign of American Fortresses and Liberators which would be flying in a "box" consisting of three columns of 250 aircraft each.

Surely we couldn't possibly miss such a "tiny" formation unless, due to my decision to "adjust" the course, we might have passed Hamburg some 70 to 100 miles to the left?

Nevertheless, I continued to strain my eyes even harder. Finally, in front of me, on the horizon, I saw some black spots. I carefully checked whether, by chance, these were not small oil drops on my windscreen. But no, I was not mistaken.

"Hello Carmen, Rainbow and Bad Hat aircraft. A large formation of planes straight ahead"'

But as we flew closer, the mirage dissolved, like a soap bubble. There were no planes, but a heavy concentration of black smoke puffs from the German anti-aircraft salvos.

I was really getting nervous. I looked at my watch, it was exactly 12.00 hours – the time of the rendezvous. They had to be somewhere there otherwise why the anti-aircraft gunfire?

By that time they must have completed their bombing run over the target, and should be on their way back! As yet I could see nothing. Where were they?

One more time I "hit" the radio. "Oxo here. Look out for the Fortress formation. They have to be in front of us! Scan the skies carefully."

We flew straight through the artillery barrage and they let us have it! I saw black smoke exploding all around us – above, below, on the left and on the right. I was getting very hot!

Then I yelled in the microphone again (although my mouth was full of saliva, and my message probably came out like confused bubbling), "Oxo here. Everybody make 45 degree right turn, the Fortresses are to our right at 14.00 hours. Watch out for German fighters in the sun! OK boys?"

They were executing a right turn in a wide arc – that great mass of silver, four-engined giants, looked very imposing. It was when the sunrays reflected on their silver bodies that I spotted them first.

Right away I felt 10 years younger. A few Messerschmitts or Focke-Wulfs would make our day. Unfortunately, the whole sky was full of our own fighters. We then began the process of taking over from the fighter squadrons, which escorted the bombers to the target.

Their task was completed, so they started back, on the way to their respective bases.

Their job might have included additional tasks and if the weather on their operational altitude was good, they would fly high, possibly looking out for some air encounters with the enemy.

Other times they would sweep down to the ground level and attack targets of opportunity such as enemy airfields, railway yards, military columns or factories. Hopefully they would return home safe and sound.

In the meantime, we had begun to take our proper positions, approaching the "box" very carefully, for we did not want to be shot at by nervous American air gunners.

Any single engined aircraft, friendly or otherwise, approaching them from the rear, was immediately fired upon.

"Oxo Leader here – Carmen, Rainbow, Bad Hat – take your assigned positions in relation to the Fortresses and the Liberators."

Each squadron took their respective positions on the right side of and above the column, echelon starboard to the rear.

From their gun turrets American air gunners were greeting us with "thumbs up" signs. We could see their faces peering at our Mustangs with looks bordering on worship. They certainly appreciated us!

By that time I was not worrying about navigation, letting the bombers' Commander take care of that detail. Our job was to protect the bomber group from the German fighters' attacks.

I scanned the skies for any sign of their presence, hoping for an opportunity to shoot one or more down.

However, all was well, there was no sign of any Luftwaffe aircraft.

In the meantime the bomber formation, probably to avoid some known anti-aircraft nests, made some course adjustments.

I checked my watch and was surprised how quickly the time had passed. Naturally, when there are no worries time passes quickly.

Since the return route had a slightly different bearing we were probably over France already.

And since the Germans did not do it over their own territory it was very unlikely that we would be attacked by German fighters over France, although one could never tell. Sometimes they were known to surprise us over our own coastline.

The group began its descent so we had to be over the English Channel.

I checked my watch again. We would be landing in about half an hour. As soon as they entered the cloud cover again all the fighter groups disengaged.

"Attention! Oxo Leader here. All aircraft make a left turn and each squadron re-form into parade formation."

We began to form tight formations before re-entering the clouds.

"Hello! Oxo Leader here. Begin descent through the cloud cover, with one mile spacing between each squadron."

In a minute we were enveloped by thick white clouds. I decided to give it more gas, since I had had enough of the sitting around and of that strange, somewhat confusing, flight.

I felt my nerves relaxing though by that time I felt very tired. I was anxious to be safe on the ground.

I saw buildings through a break in the clouds, dived through it and then in a low flight approached our airfield.

"Prepare for landing."

Everybody knew what to do. My Mustang landed hard on the steel net and continued to roll to the end of the runway. We were back!

Passing the dispersal tent I heard Dziubek's voice grumbling: "How the hell did he find the target? That I don't understand! Let's go to the de-briefing, maybe we will learn something!"

The Operations Room was full of noise. Almost everybody was there.

Immediately I assaulted our Intelligence Officers with questions, but they would not let me speak. Everyone was talking at the same time.

Finally, I silenced them, so that I could address the Intelligence Officer.

"Why don't you tell me where I would end had I stuck

to the course that you gave me? Go on – tell me! Why don't you re-calculate?"

"But let me explain old man . . "

"Don't say anything! Just re-calculate!"

With reluctance the Intelligence Officer began to re-calculate our original route.

Something certainly did not fit! "Sometimes following the tip of one's nose is better than a badly calculated course" said F/Lt Potocki, illustrating his statement by pointing to his own outsized schnozzle.

S/Ldr Haw, the Englishman, addressed me: "I really don't understand anything about that flight! What do you say to that!"

"I really don't understand it myself, what made me adjust the designated course the way I did!" I replied.

"I have re-calculated the course!" interrupted F/O Przymienski, the Intelligence Officer.

We all looked at the map. "Just as I thought! If I stuck to the course we would have drifted towards Magdeburg."

"May I explain something?" asked F/O Przymienski.

"Please do!"

"When you were approaching The Hague I received a new meteorogical forecast. It turned out that the wind at your altitude had suddenly changed by 90 degrees and the original course was calculated with the headwind."

We all looked at one another. I felt that Haw and Dziubek had forgiven me.

"Having the latest information, and realising the seriousness of the situation, we calculated the new course. Then we called Operations Room to give you new data. Unfortunately, you were already out of range. Could you hear anything?" asked F/O Przymienski.

"Nothing!"

Stanislaw Skalski, W/Cdr 76710.

In June 1947 Skalski voluntarily returned to Poland, continuing his service in the Polish Air Force which was under Communist rule.

He was arrested for espionage in June 1948, tried and condemned to death. As a result of his mother interceding his sentence was commuted to life imprisonment.

Following a review Skalski was released in April 1956, having served 8 years in jail. He continued his Air Force career, retiring as General.

Skalski returned to Coolham from Poland in June 1994 attending the commemorative weekend as guest of honour.

Ground crews attend to W/Cdr Skalski.
Photo: J. Polak

F/Sgt Jozef Korczowski (left), chief mechanic 315 Squadron and W/Cdr Skalski.
Photo: J. Polak

4

315 Deblinski (Polish) Fighter Squadron

Squadron Leader Michal Cwynar recalls that on a dull afternoon, 1 April 1944, 306 and 315 Polish Fighter Squadrons took off from RAF Heston for the newly built landing strip at Coolham.

Their first few weeks' flying from Coolham were spent completing their conversion from Spitfires to P-51 Mustangs which he said were "wonderful aircraft to fly".

It was going to be, as later events proved, three months of exhilerating and sad moments while flying from this base, before and during the invasion of Europe.

Coming in to land on a new landing strip, covered with Sommerfeld metal mesh and with unfamiliar landmarks surrounding the aerodrome, one always took great care with the approach. The idea was to touch down as close as possible to the beginning of the runway to avoid any possibility of overrunning.

I executed a left hand turn, undercarriage and flaps down, just skirting the east side of Coolham and, when touching down, heard a terrible rumbling noise as if my Mustang wheels were running on a cattle grid!

To soften the outrun after landing we often used, albeit gently, foot brakes. I did not dare touch the brakes this time because I was worried about the undercarriage tyres.

Our squadron mechanics, already on the airfield, beckoned us to taxi towards the west side area of the north-south landing strip to a designated dispersal point alongside large trees.

I stopped at the point indicated by my mechanic, switched the engine off and immediately jumped out of the Mustang and went to examine the aircraft's undercarriage wheels but the tyres were perfectly normal.

We gradually got used to taking off and landing on the metal mesh covered runways and thought no more of it. We took off and landed in sections of two aircraft, just as on tarmac runways.

During the first few weeks of flying from Coolham we were still in the process of completing our conversion from the Spitfire VB to the P-51b Mustang with original Allison engines and those awful Messerschmitt-type canopies, which were soon phased out, replaced with Packard-Merlin engines and Malcolm bubble canopies. They were wonderful aircraft to fly.

Our 315 Squadron was led by S/Ldr Eugeniusz Horbaczewski, a demanding man of his pilots during flying exercises as he put us "through our paces" and, as always, made us push the Mustang's performance to its limit.

There was no more shuddering fuselage and wings, Spitfire style, when over 400mph, yet, in certain circumstances, the Mustang could be unforgiving, but once you got that certain indescribable "feel" of it you had complete confidence and joy in flying this wonderful fighting machine. There was an extra bonus – comfort and a first class heating system!

Our C/O's approach to training flights was imaginative, whether dogfighting, or dive-bombing – new to fighter pilots and completely different from strafing ground targets.

First Dziubek (our C/O's nickname meaning Little Beak) went through the pre-flight briefings on the ground, in front of a blackboard showing 12 aircraft in close echelon starboard, flicking and diving, dropping bombs from a certain height and angle to predict trajectory and increase accuracy. This procedure was then put to the test during flying practice then back to the airfield and the blackboard for de-briefing. Great fun!

On one occasion, Dziubek, leading the whole squadron, was trying to figure out what would be the best tactics to engage a large, squadron size formation after a sudden, head on interception. Here, the analytical mind of "A" Flight commander, Henryk Stefankiewicz, came to the fore.

Henryk's theory was as follows: From the battle formation of three sections, each made up of four Mustangs, the leader's section and the outer right section should make a sudden climbing turn to the left while the inner section should commence an unorthodox manoeuvre, climbing through eliptical loops in the opposite direction.

The enemy's leader, forced to make a left climbing turn to engage our leader's section, would leave the rest of his flight exposed to our inner section exploiting the flying space above. Dziubek thought it was, in principle, a very good idea.

On another occasion, Henryk suggested to Dziubek that three of us should, at a safe height, try to simulate a dogfight on the "deck" i.e. at ground or sea level.

When flying low, the only effective manoeuvre at one's disposal was a steady, tight turn. When, after four circles, there seemed to be a stalemate, or the adversary was gaining, the suggested idea, which I had applied many times during dogfight practice on Spitfires, was to lower 10 degrees of flap when the speed dropped to about 230mph. It always proved successful.

So, with Horbaczewski flying above us as observer, Henryk and I approached each other at an agreed speed of around 400mph. When passing on the customary port side, we engaged in a left hand turn dogfight. After the two 360 degree circles, the speed dropped to about 230mph and the pilot who, at this stage, lowered the flaps to 10 degrees, was always gaining the advantage. But there was a considerable danger in this. With 10 degrees of flaps, attempting to pull too tight a turn, the Mustang could suddenly flick and stall.

Back on the ground we agreed that it was an effective ploy but Horbaczewski thought it too dangerous to advocate openly, saying: "I am not having my pilots practising dogfights with flaps down and falling to the ground like kites from Coolham skies. That's the end of this experiment – officially!"

But after we had left Coolham for good, six of 315's Mustangs, led by Horbaczewski and without a single loss to themselves, shot down seven Messerschmitt Bf109's in a dogfight over Norwegian Fjords. Ten degrees of lowered flaps came in useful. Somewhere, in the Air Ministry archive dungeons, is a film of this action, made by myself.

While still converting from Spitfires to Mustangs at Coolham during April 1944 one of our squadron's pilots – tall, athletically built and with a hardly noticeable limp – had great difficulty landing his Mustang.

Despite being an experienced Spitfire pilot, before his final, critical, three-point landing, his imaginary ground level always seemed to be one or two metres above the actual. As a result of this misjudgment his Mustang, already at critical stalling speed, thumped on the runway.

On one occasion, standing beside Horbaczewski, outside our dispersal tent, we watched four of our 315 Mustangs coming in to land. As this particular one was about to touch down Horbaczewski turned his head in the opposite direction, pretending to watch trees, clouds, anything but this pilot landing.

Laughing, I said: "It's all right, Dziubek, he is safely down."

The C/O turned back and replied: "All right this time – until next time!"

To end the saga I must digress and refer to the end of June when for a short time we were flying from RAF Ford. After taking off on a long range mission with droppable fuel tanks we were recalled after only about 20 minutes in the air when the mission was cancelled, our droppable tanks still completely full.

At Ford, landing last, this pilot hit the ground with the left wheel. The whole of the left undercarriage and the droppable tank sheered off. The Mustang veered in a long left hand semi-circle, losing around half of the left wing before coming to a stop.

With the horrific and tragic death of F/O Szajda at Swindon just a few months before still on our minds we watched to see if the spilled fuel from the sheered droppable tanks would ignite as it had done with Szajda.

Luckily, probably because of the sandy soil at Ford, there were no sparks on touch down and no ignition.

Horbaczewski's patience snapped. Next day, the pilot was packing his belongings, posted to different pastures. I felt sorry for him this time. It was just bad luck hitting a bumpy surface with the left wheel. Coolham's mesh runways were much superior.

I felt like pleading his case with the C/O but it was too late, the die was cast.

Yet later, Horbaczewski felt uneasy. That night, in our tent, he said to me: "Next time some clever dick from Ops room tells me to land with full droppable tanks, after making sure that all my pilots are flying on fuselage tanks, and while still over the Channel, I will order them to jettison the bloody droppable tanks into the water before landing anywhere!"

Describing the process of converting to Mustangs has taken my resumé of our sojourn in Coolham too far ahead of the first few weeks' mundane existence on the edge of that pleasant Sussex village.

Life was spartan at Coolham airfield. I found myself going from the comforts of Heston vicarage, where I had previously been billeted, to living under canvas for three months in tents. The first few weeks of April the weather was damp and cold with frequent rain and muddy tracks around the tents.

By having campbeds and comfortable sleeping bags we pilots were lucky. Ground crews in their tents slept on straw, rough made mattresses and had a few blankets to keep warm, that was all.

My tent companion in Coolham, as for the previous three years on other RAF stations' living quarters, was my close friend F/O Adam Sworniowski. He was an idealist and perfectionist whether it was his personal appearance or manners and he was a pilot of great natural ability and talent.

Apart from a personal, mutual affinity, our friendship was, in a way, complementary. He emulated my flying technique while I, in close association with him, had the chance to improve my fragmented knowledge in other areas related to flying and in general subjects.

On the east side of our tents, about a hundred yards from us, was a little brick building called, grandly, the officers' bar.

Adam and I never frequented that craggy place but could hear, during those cold, rainy April nights, exuberant voices coming out from this bar – our colleagues singing Polish songs.

Our dear friend and, at that time, our "B" Flight commander, Janusz Marciniak, acted during those singing sessions as a self-appointed choirmaster, but a choirmaster with a difference!

After having a few drinks, Janusz picked one or other reveller, accusing them of not singing in harmony. At that point, the worst thing the accused could have done was to admit to it, saying, "Mea Culpa", because that was quite the opposite reaction F/Lt Marciniak expected. He was just trying to create a little controversy.

Sometimes, as Adam and I learnt later, Janusz got more than he bargained for!

In 306 Squadron there were a few pilots with short fuses. We knew one who, when unable to articulate his point of view, was inclined to resort to other means.

Little did these pilots know, however, that in just a few weeks' time dear colleague Janusz Marciniak, a solid, loyal friend but a strict disciplinarian when it came to duty whether flying or on terra firma, was to become Commanding Officer of 306 Squadron.

Adam and I often wondered if there were any repercussions, knowing that Janusz did not suffer fools gladly!

About the middle of May 1944 our reaction to an hilarious incident was to quote a Chinese proverb – "We found ourselves between tears and laughter."

At that point during our stay at Coolham our tented living quarters were still near a side road leading to Coolham village.

Early one afternoon our pilot colleague, F/O Kazimierz Sztramko, ran towards our tent calling on Adam and me to come quickly out on to the road to see what had happened.

It transpired that the Administrative Flying Officer of 131 Wing was supervising a party of Polish airmen, emptying Elsanol containers into large, empty oil drums placed on the platform of a lorry. The contents were to be taken to, and emptied into, a prepared septic pit somewhere on the east side of Coolham village.

When the lorry was loaded and ready to go, the supervising officer, a scrupulous fellow, instead of taking a seat in the cabin beside the driver, decided to show everyone who was in charge of this dangerous expedition.

So he climbed on to the back of the lorry and stood just behind the driver's cabin, holding on to the upright board, facing forward, as they drove through the village.

Going through Coolham, for some reason or other, the Polish driver slammed on the brakes.

On the back, the officer in charge first jerked forwards, then, through the whiplash effect, dived backwards, straight into one of the large oil drums!

The poor fellow was fully submerged in the "elixir of life" and had to be pulled out of the drum. He looked like a bar of chocolate.

By the time we arrived on the scene, the officer in charge had already suffered the indignity of being stripped out of his complete attire, was wrapped in blankets and being loaded into another vehicle to be driven to hospital in Horsham.

There, he was disinfected and bathed and was given an anti-tetanus injection.

Meantime, the task of driving the lorry and its contents to its destination had to be completed but the airman in charge, wise chap, decided to proceed on foot along with his colleagues to avoid the unpleasant downdraft from the precious cargo that was all over the front of the lorry.

About the middle of May, 315 Squadron's living quarters were moved to the south west side of Coolham's landing strip. A large marquee was erected nearby where the following evening an ENSA concert party from London arrived to entertain 131 Wing personnel.

Among the artistes there was a pretty soprano, accompanied by a pianist, who sang popular wartime songs and operettas.

Horbaczewski, sitting in the front row wearing his best suit – he never wore battledress – made up his mind to meet this star at the after-show reception. His enduring love affair had begun.

One week before the invasion of Europe, late on a Sunday afternoon, our C/O, Dziubek, asked "A" Flight's Henryk Stefankiewicz and me to put our best suits on as he was going to drive us to Brighton in his jeep.

We pulled up outside an imposing hotel (the Grand Hotel) on the promenade, to discover that our dear colleague, former "B" Flight commander of 315, now C/O of 306 Squadron, Janusz Marciniak, had invited his three best friends from 315 to meet his bride, Emily, and entertain them to dinner.

After living rough in tents for the past two months walking on a luxuriously carpeted hall to a lounge and dining room felt like being on a trampoline.

We spent a pleasant evening with the happy couple and then drove back to Coolham. I remember Mysza – as Stefankiewicz was affectionately called – was anxious about his spaniel dog being hungry and waiting for him to feed him, while Horbaczewski, driving the jeep, reverted to silence.

Mercifully, I had no premonition that during the first week of the invasion of Europe, Henryk would be shot down by flak near Cherbourg and killed; that during the same flight, our C/O would perform a daring rescue of another pilot, also shot down by flak; that our dear friend, Janusz Marciniak, who had entertained us that very evening would, two days after we had left friendly Coolham for good, also be shot down by flak and killed, leaving his charming wife whom we had just met, a young widow.

Worst of all, that two months after leaving Coolham, flying out of Brenzett, a Polish padre would be coming into the tent I shared with Horbaczewski, to discuss with me how to break the tragic news to that pretty soprano, now living in London, that our C/O – Horbaczewski – had been killed in action in Northern France.

On 29 May, late afternoon, Air Marshal Coningham, commander of 84th Tactical Group, flew into Coolham, inspected the squadrons, then took all the pilots of 131 Wing to the centre of the airfield.

He asked us to form a circle around him, so everyone could hear, and told us that the invasion of Normandy was soon to commence.

He briefly explained what functions our squadrons would perform, shook hands with all of us and wished us good luck.

After the briefing we were asked to retire to our tents and were told not to speak to anyone about the briefing.

On 6 June 1944, it was a dull day and 131 Wing escorted a large formation of gliders being towed to Normandy beaches, north of Caen, i.e. the British troops' landing section.

On 7 June, flying to the landing area and beyond on armed reccies in groups of four Mustangs, 315 Squadron made five sorties.

On 8 June, after an early morning briefing, I went to have a word with F/Sgt Jozef Korczowski, our squadron's chief mechanic, to inquire about my "PK-Z" Mustang and was told that this aircraft would not be ready for flying that day.

Our C/O, Horbaczewski, came on the telephone and in the presence of Korczowski asked me to take command of our "B" Flight and to be his deputy.

Our chief mechanic, a giant of a man with powerful arms, grabbed me under my shoulders, whirling around several times, emulating a Polish folk dance. Korczowski and I were very fond of each other.

That day, I was going to fly a Mustang "PK-U".

Although I was not superstitious, I was never fond of flying aircraft with the letters "U" and "Y". When flying operationally with these letters, usually, something went wrong.

But nevertheless, on this day, after coming from the ranks where I had been since 315 was formed, to be a flight commander and Horbaczewski's deputy, I was on the top of the world.

Later, at 15.20, leading a Flight of four Mustangs, I took off. We were armed with 500lb bombs under each wing and we headed out to the Falais–Alencon–Vire area to dive-bomb certain targets and later to patrol as fighter protection over the Allied landing zone.

Approaching Caen at 2,000 feet I heard a loud bang under my seat somewhere near the air cooling intake on the fuselage.

My wingman and trusted colleague, W/O T. Jankowski, a good raconteur when on the ground, called me on the radio transmitter (RT) and said: "Michal, your Mustang is belching grey smoke. You've been hit by ground fire."

Turning back, I ordered Jankowski to take charge of the remaining section and proceed with the designated mission towards Falais–Alencon–Vire.

Preoccupied with jettisoning bombs and deciding where to make a forced landing I was not aware that my colleagues circled above for a while to see if I would get down safely.

My engine lost power completely – it had probably seized – and the propeller blades were turning only by force of the aircraft's speed. I tried to reach the Allied beach perimeter but I was losing height rapidly. I fought to maintain a gliding speed at around 160–170mph and attempted to lower the aircraft's flaps but without success.

The hydraulics must have been shot-up so I decided to land on a large, lush cornfield.

I touched down smoothly enough with the undercarriage up and it was only then I noticed high wooden posts coming towards me and whizzing by. These posts were anti-glider obstructions erected by the Germans.

Inevitably, the tip of my Mustang's left wing hit one of the posts skidding the plane round in what was almost a 360 degree turn, before coming to a halt.

When I got out of the Mustang I saw lots of explosions and learned later they were from a major tank battle in progress with the Allies advancing on the German lines.

Running in the opposite direction to the fire blazing area, through the knee-high corn and in a north-west direction towards the afternoon sun, I was nearing the end of the cornfield when I heard a loud voice calling in English: "Stop running! Stop! The field is full of landmines!"

I got the message and gingerly began to tiptoe towards the voice. It belonged to a British corporal who drove on his motorcycle to pick me up.

Later, I learnt from him that his task was to rescue Allied pilots who had been shot down and to drive them to the safety of the British Forces' beachhead.

On the coast, among the dunes, I was driven to a small tent and my rescuer led me inside, pointed to a campbed, told me to rest and that he would be back in about two to three hours to drive me to the rescue speedboat.

Resting on the bed I heard lots of activity outside, soldiers' voices shouting and cursing and I got the impression they must have belonged to the Royal Engineers Corps.

I also heard the voices of young English officers, with exaggerated public school accents, ordering men about.

In about 20 minutes or so a tremendous barrage opened up from British naval ships, lying off shore.

I went out of the tent and saw in the sky to the east, what must have been two squadrons of FW190's, lining up to strafe British positions.

Soldiers, cursing, kept running for cover under armoured vehicles and dug-outs among the dunes. There were no public school accents from the equally scared young officers caught up in the action. Having lost their nerve they dived for cover as well. Fear is a great leveller!

I went back inside my tent and, ignoring the campbed, lay on the ground, face down, arms covering my head, and thought – what rotten luck!

About this time I should have been flying back from the Falais–Vire area and in position to engage German fighters. For the first time they were over Allied territory, albeit small.

In all these years of flying over the Channel and enemy territory to engage German fighters, we pilots knew, but never spoke about, that from the start the enemy pilots always had the psychological advantage, knowing full well that if they were shot down in combat over their territory they could force land or bale out and still be safe. We, British and Poles, after engagements, had still to cross the Channel to the safety of the English mainland. After about three hours my rescuer, the motorcycle rider, returned to the tent and informed me he was going to drive me to the beach to a British navy motorboat.

There, he carried me through tidal water to a sort of wooden platform, a jetty, and with the help of sailors, got me on board. During this whole episode, from rescue in the mined cornfield to this moment of tender help, this British corporal was exceptionally calm and composed.

I shook his hand firmly, thanking him for taking care of me and we wished each other good luck.

Unfortunately, the sea during the return journey to England was very choppy and rough.

As the speedboat skimmed over the waves I lay on a wooden bench, covered with a blanket, feeling awful.

A sailor gave me some rum and after gulping it down I began to feel very warm not having had anything to eat since the previous lunchtime. I found the rum had an immediate affect on me.

I felt that the bench I was resting on was making slow rolls. With both hands I held grimly on not to fall off it.

Unable to sleep, the overnight sail seemed to last forever. At dawn, approaching coastal waters, the sea was calmer and I felt a little better.

Shortly afterwards we arrived at Calshot naval base and I was driven to the officers' mess where my C/O from Coolham, Horbaczewski, was already waiting for me.

He must have got up quite early in the morning to drive down from Coolham to collect me. We embraced warmly.

After breakfast Horbaczewski drove back to our landing strip and as they say, "I was back in business".

When we returned to Coolham, F/Sgt Korczowski, our chief mechanic, told me that my "PK-Z" was serviceable for operational flying.

On the third day after my return from Normandy, after the early morning briefing, Horbaczewski announced he was going to lead the first operational sortie. Then he took me aside and informed me that HQ Fighter Command had informed him that a telegram had been sent to my wife in Dumfries saying that I was "missing".

"Take a Mustang now and fly to Dumfries to show your face and prove that you are safe and well," Horbaczewski said.

So within a few minutes on that sunny morning, I took off, landing at RAF Heathhall, Dumfries, in two hours.

After lunch with my family I took off from Heathhall, landing back at Coolham at four o'clock in the afternoon.

During those momentous first few weeks of the invasion of Normandy our 315 Squadron flew at least two sorties each day. So the same evening of my return flight from Dumfries, at 18.00 hours, another briefing for the last operational flight of the day was held.

Outside the dispersal tent, Horbaczewski, standing beside his blackboard, started to write the names of pilots to fly on this mission.

First, to lead this sortie, was Michal Cwynar, followed by those of Adam Sworniowski, Tadek Slon and Tadeusz Jankowski. That morning, Horbaczewski could easily have instructed a clerk in Fighter Command HQ to send another telegram to Dumfries, notifying my wife of my safe return to Coolham. Instead, our C/O had generously suggested that I should fly to see my wife but now, to show the rest of the squadron's pilots that nobody was favoured, he designated me to lead the last mission of the day.

Demanding yet generous and, most importantly, always fair. That was why all the pilots respected him so much.

On the morning of 10 June, at 6.45, my close friend and tent companion, F/O Adam Sworniowski, led a section of four Mustangs, with bombs, on armed recce to the vicinity of Falais–Vire.

South of Caen he was hit by enemy flak. He jettisoned his bombs and crash landed safely on territory still held by Germans.

Adam's wingman saw him getting out of the Mustang and start to run towards nearby trees for cover. Then they saw him suddenly falling to the ground, shot and killed by enemy small arms fire on the ground.

It was a terrible blow for me, for all of us in 315 Squadron. I lost a dear friend, a man of great courage and exemplary integrity. Had he lived I am sure he would have reached the position of a squadron commander.

A few days afterwards his cousin, who was a navigator in a Polish bomber squadron, came to Coolham to collect his personal belongings and we promised to keep in touch.

For the rest of our sojourn in Coolham I was in the tent alone. I missed Adam and still do! Every time I hear actor and comedian Harry Secombe's high pitched voice Adam appears in my mind's eye.

On 22 June, 306 and 315 Squadrons took off for the last time from Coolham, leaving the friendly village we had become so much attached to with both pleasant and sad memories while flying away from there.

In 20 minutes we landed on a sprawling landing strip called Holmsley South.

Two days after leaving friendly Coolham our dear colleague, 306 Squadron commander Janusz Marciniak, while on an operational mission, was shot down by flak and killed.

My recollections describe only a few fragments of the activities that had taken place near the West Sussex village of Coolham 50 years ago.

No doubt other surviving personnel of 131 Fighter Wing will have their own memories to relate.

Mr Paul Hamlin literally rediscovered the historical importance of that wartime landing strip, travelling in this country and abroad, undertaking painstaking research to build a true picture of activities from that base in 1944.

Thanks to Paul's noble efforts, the history of those momentous times will be preserved for future generations and posterity.

Michal Cwynar, DFC, F/Lt P-1903
Scotland.

❐❐❐❐

One incident when one of 315's pilots was killed left an indelible impression on Flying Officer Bozydar Nowosielski-Slepowron who looked up his log book entry for 19 April 1944, which read: P/O Brygider broke up and died.

I remember that flight well. Mysza (Stefankiewicz) took the squadron up for "Battle Formation".

We had had the Mustang for less than two weeks, as I had my first solo in the Mustang on the 3rd, in Coolham, and on the 5th I brought Mustang FB179 from Aston Down via Ford. I remember that three of us went to fetch the planes. Went in Spits and came back with Mustangs.

P/O Brygider was my No. 2 and we were in Mysza's four. Mysza dived to about 50 degrees, simulating a strafing attack, so his speed went well over 470mph (505mph was the maximum allowed for Mustang P-51b at that time) and the squadron tried to keep up.

When my speedo showed over 500mph I cut down the power and started to pull out of the dive very gently. Brygider was tougher, he passed me increasing his revs and – his tail came off!

After that, bit of panic, those following me also pulled up and the squadron landed in bits. I was the last to land as I got very scared. Every time I put down the flaps, I went around, because I noticed that the trailing edges of the wings folded down. This is perfectly normal for the Mustang, but in the stress of the situation I forgot it, being used to the split flap of the Spitfire.

It took me three times before I landed. We were grounded for two or three days until all the planes were inspected and a special plate about 8 x 4 inches was riveted to the offending part of the tailplane.

Next time I went up was on the 23rd with Horbaczewski leading the "Battle Formation". The squadron was operational on Mustangs on 1.5.1944 – Ramrod 818 over Charleroi!

Bozydar Nowosielski-Slepowron, F/O P-136564
Lydney, England.

❑❑❑❑

Pilot Officer Bogdan Smidowicz adds, regarding P/O Brygider:

"I flew from Coolham between 11 April and 28 May 1944. I was the 315 Squadron Adjutant and among many matters I also had the sad duty to bury my colleague, P/O Brygider, who lost his life striking the ground near Coolham airfield when diving in his Mustang."

Bogdan Smidowicz, P/O P-2340
Gdynia, Poland.

❑❑❑❑

An incident that could have ended disastrously but had a funny side that amused the airmen afterwards is related by Flight Lieutenant Jerzy Polak.

Coolham airfield was surrounded by mountains of high octane cans of fuel.

One of our pilots damaged a doodlebug (V1 flying bomb) in the air and it came over the airfield flying all over the place, so all the ground and flying crew were running away from it.

It makes me laugh, the way we were running like chickens.

Fortunately, it went down a few hundred yards from the perimeter and exploded. No-one was hurt.

Jerzy Polak, F/Lt P-1802
Blackpool.

❑❑❑❑

Another 315 Squadron pilot, Warrant Officer Tadeusz Slon, said that at the beginning of April 1944 they were on Spitfire V's and then the Mustangs started coming in when they transferred to Coolham where they found conditions primitive. His youthful exuberance also landed him in trouble on more than one occasion.

There were no living quarters as such at Coolham. You had a tent and a campbed. It was a wooden contraption with canvas on top, always collapsing.

Eventually we had to collect hardboard boxes and sleep on them. We were sleeping really on a hardboard floor.

And one time one of the pilots breaking up the box with rusty clips got poisoned blood and his hand swelled up and he had to have a tetanus injection.

There was a lot of rain and if you touched the tent from the inside they used to leak.

Food, at the beginning, was government issue, almost nothing, but later on it was decided they would feed the local pigs and we had as much bacon as we wanted including Polish sausage.

For entertainment, there wasn't any apart from the local pub but at the weekend I used to walk to Billingshurst with a Flight Sergeant from 129 squadron and we used to go for a pint.

On the way from Coolham to Billingshurst, on the side of the road, there were stacks of the 500lb bombs the Mustangs used to be fitted with all the way to Billingshurst.

Regarding training, before the squadron could be operational everyone had to do at least five hours' flying time.

Horbaczewski, our C/O, said everybody must do four hours and four landings as quickly as possible because he was itching to go in.

Eventually, on 29 April, we were joined by 129 and the Wing – the three squadrons – had its first operation against France.

It was in bad weather and because of the cloud no one saw the bombers and eventually the squadron started coming back.

But Michal Cwynar – I was in his section – when we got ordered to go back, he said, "ignore it" so we broke cloud about a couple of hundred feet above the ground over France and we spotted an aerodrome which must have been bombed the night before and some civilians were filling up the craters.

We were shooting at them and today, when I think of it, I think it was probably forced labour, filling the holes for the Germans.

Anyway, we landed back in Coolham and they said Horbaczewski had been taken to hospital because when he was coming back shrapnel hit his hood and he had a cut on his face from perspex.

An accident I was involved in during training occurred when there was an ambulance on the airfield. We were in each other's way and I was taxiing and we touched. The ambulance being plywood just disappeared and I was sitting like a wally saying bloody hell!

Another incident was when I had the hood open on the aircraft (it should have been closed), taxiing. I turned left, I had to open the throttle, the gas, and the bloody thing just flew off. Being sideways it just flew off.

F/Sgt Korczowski (chief mechanic to 315 Squadron) came running over and said: "You son of a bitch, that's your second one," and a couple of days later I had my third one, low flying, so he had had enough of me.

Eventually I came to the conclusion I would have to control myself.

The low flying incident was with an American Thunderbolt aircraft. I took off to do my last tour of the day, fourth or fifth hour on, and I spotted a Thunderbolt flying near Coolham and I thought what is he doing here, bloody Yank. So we had a scrap, sparring as if in a dogfight.

Unfortunately, this had been witnessed by others on the ground at Coolham. When everything was over and I landed someone ran over and I was told to switch off the engine and stay there "because the station commander is sending the police to arrest you".

So I switched off, went in the ditch, smoking a cigarette and shaking and eventually, about one hour later, a fellow ran out and said to taxi back to the proper position.

With shaking legs I went in and said: "Sorry sir, I don't know what came over me." So he said: "You know, a lot of people don't like to see anyone enjoying himself. So in future, go 30 miles away to enjoy yourself where nobody see you instead of here because you will be charged."

And he gave me a cigarette and he said "forget it" and to this day I cannot think why I got away so lightly.

But Horbaczewski didn't give a damn. He was a man who allowed his pilots to express themselves and he said: "If you want to express yourself you don't fly by rules, the rules are made to be broken."

But two weeks later the adjutant called me in and he said: "Read that." He gave me a bit of paper, I read it and it said: "Lack of discipline, Warrant Officer Slon is going to go to Aston Down as a punishment." I was told to keep quiet until Horbaczewski came.

So when Horbaczewski came he too read it, then picked up the phone (to Polish Group HQ) and said: "I thought I squared the bloody thing two weeks ago about Slon and me. If anybody's going to punish pilots I am the C/O, I am going to punish him. In future if you want to do it I am going to resign."

So words were spoken and Horbaczewski calmed down and he said to me: "You go to Aston Down and you will be back in a few weeks. But in the meantime, if ever you have a chance do drop in for a chat."

So off I went to Aston Down and the Wing Commander there said: "Ah, you're the bad boy. You know there are King's Regulations and you are not supposed to do that." I said: "I'm sorry, Sir, I don't know why I've done it, I'm young." So I asked: "What am I going to do here?"

He said: "You are going to deliver aircraft to different squadrons and probably eventually your squadron will take you back."

So I said: "Can I have a free flight?" and he said "Yes", and I took a Spitfire 14 which was one of the latest ones with five props. And I said: "Can I land at Coolham?" and he said "Yes."

I landed there and reported to Horbaczewski and I said: "Sir, I am here," and we had a good, friendly and private talk which I don't want to say too much about.

He was lamenting about the Polish authorities about how they treated us. It was all King's Regulations and the Polish ones and they used to pick two of them and for instance if the English charge was higher than the Polish they used to charge you under the King's Regulations.

He said: "Who are the bad boys? The bad boys are the good boys. How could you judge a bad boy – a good pilot who makes a low roll and he is picked up as a bad boy and disciplined. The point is he can do it. All the rest can't so they charge him, lack of discipline, disobedient."

Eventually, I got back to the squadron.

At Coolham a typical mission around the D-Day period would start differently from, say at Heston, as at Coolham we had no Met. section as such, you just had your name put on the board and the officer would say the weather is such and such, that was all. The briefing also only amounted to being told you were flying this afternoon, take off so and so.

Usually we carried bombs. Originally it was 12 Mustangs carrying bombs, there was no escort, but a couple of times squadrons were caught out and planes blown out of the sky. So later on they said eight aircraft with bombs and four as fighter escort.

306 Squadron had been twice the losers, caught as 12 bombers.

When we took off or came back to Coolham, Hastings was the entry and exit area.

We had no radio aids if the weather was bad, it was fly by eye. For navigation you had to rely on yourself from beginning to end.

For relaxation at Coolham we used to dredge the lake for eels and then, taking a wooden barrel, covering it up, made a fire underneath with fresh wood which gave out a lot of smoke, smoking the fish and after a few hours they were nice and brown.

One of them used to catch eels on a line.

I remember one Sunday morning I saw Horbaczewski and he didn't look well so I said: "What is wrong? Do you have a problem?" And he said: "I have borrowed Nowierski's (the Group Captain) car and I walloped a bloody lamppost and done damage."

So I said, "Stay here", ran to the airmen's tent and said: "Boys, the boss is in trouble." They all rushed out and almost carried the car under cover. In two hours they beat the bumper up and they said if possible they would deliver the car tomorrow when the paint would be dry.

So Horbaczewski looked at me and said "thank you" and next day he told me: "I've delivered the car. He never even noticed it."

So that's how we helped our boss. He didn't ask for it, it was our spontaneous idea to help him.

After we left Coolham I can clearly remember Dziubek's (Horbaczewski) final day on 18 August 1944 – the flight to Beauvais.

The day before he was ill but not that ill and he said: "Tomorrow morning I will decide who is going to lead the squadron." Pietrak, his deputy, would have taken over. But on the 18th Dziubek was OK (although in another account Dziubek is said to be still quite ill). He rubbed his hands and said: "I feel I'm going to have a scrap."

The briefing was very, very short, but he told us there would be 12 of us, "no break-offs, stick together". He said: "I am responsible for you, I don't want to write letters to your mothers."

It was a beautiful day, sunshine, quiet, peaceful. We were flying at 8,000 feet and we were looking for German aircraft to attack us, looking for trouble, around Paris, from the west. If nothing happened there, turn north, go down and strafe ground targets.

I saw Germans taking off, some circling round already and when Horbaczewski attacked them he attacked those which were airborne.

I told him about the group. I saw it after passing the aerodrome because I was the tail end Charlie. My duty was to look and cover the front ones and myself.

I once asked him why he always chose me as tail man and he said: "Because I trust you," so he shut me up.

Anyway, this day it developed into a dogfight, a scrap. There were a lot of fires around. There was silence. Nobody talked. Dziubek didn't talk and eventually Pietrak said: "Dziubek?" No answer. Again, he repeated "Dziubek" and again no answer so Pietrak said he would take over.

We congregated at 10,000 feet over Beauvais and they said: "Who is missing?" and, of course, "PK-K", Dziubek's plane, was missing and Pietrak said: "Right, we are going back."

We said he may be shot down, hiding. Nobody thought he would be killed.

Eventually 1945 came and there was still no news of him. Our staff officers were searching after the war, everywhere, and could find nothing.

His mother found her son's grave in 1956 in Creil due to help from the French Red Cross and the German equivalent of the British Legion and our so called people, his friends, found nothing.

Tadeusz Slon, W/O P-781269
London

❑❑❑❑

A fitter with 315 was Corporal Peter Skrzypczak who said:

My memories of the war years are not as vivid as some of the chaps.

All I know is that my time at Coolham was hectic and exciting. What with new aeroplanes, casualties and repairs to be attended to the ground crews in the field were very busy from early morn till late at night (well, most of the time).

Peter Skrzypczak, Cpl P-784044
Southampton.

❑❑❑❑

Sergeant Stanislaw Dzialowski, was a fitter with 315 Squadron on "A" Flight and recalls:

I was the only person left working on a Spitfire in the squadron when a new Mustang arrived which I was put in charge of.

It was delivered by a ferry pilot whose face was badly disfigured by burns. He had an artificial leg and arm and I was very surprised at the state of the pilot as he had to ask me to help him out of the cockpit.

He told me to get the Spitfire ready for take off and as he took off he crashed into the trees but luckily he lived.

When the new Mustang pilot was introduced to me he asked me if I could paint him a mascot onto the plane so on the left hand side of the engine I painted a stork carrying a baby.

This was the aircraft that landed in the marshes of France from which W/O Tamowicz was rescued by S/Ldr Horbaczewski.

In the squadron was a car that had been in a crash and I bought it for £1 and named it Doodlebug. Everyone knew about me and this car.

Stanislaw Dzialowski, Cpl P-784337
Grimsby, South Humberside

Sgt Stanislaw Dzialowski

315 (P O L I S H) S Q U A D R O N
RAF Llanbedr, March 1944

Back row, from left: F/Sgt J Bargielowski, F/O B Nowosielski, F/Sgt K Siwek, F/Sgt B Czerwinski, F/Sgt J Gozdzielewski, F/Sgt A Ciundziewicki, F/Sgt J Adamiak.
Centre: F/O P Kirste, F/O P Kliman, F/Lt M Cwynar, W/O J Derma, F/Sgt T Slon, F/Sgt S Bedkowski, F/Sgt Luzinski, F/Sgt T Berka, F/Lt R Idrian.
Seated: F/O B Smidowicz, F/Lt A Sworniowski, F/Lt H Stefankiewicz, F/Lt M Gorzula, S/Ldr E Horbaczewski, F/Lt J Marciniak, F/Lt J Polak, F/Sgt T Jankowski.

Photo: M Cwynar

Pilots of 315 Squadron gather round S/Ldr Horbaczewski. *Photo: M Cwynar*

F/Sgt Tadeusz Slon (left) with S/Ldr Eugeniusz Horbaczewski. *Photo: T Slon*

F/Lt Michal Cwynar, DFC. *Photo: M. Cwynar*

"Booze up" – F/Lt Franek Wiza of 315 Squadron, far right, with members of 129 (Mysore) Squadron. *Photo: C. Haw*

315 Squadron mechanics with puppies, offspring of cocker spaniel bitch owned by F/Lt Stefankiewicz. They were adopted by personnel after he died in action on 22 June 1944. *Photo: M. Cwynar*

Cpl Jozef Woroniecki a fitter with 315 Squadron taking a well earned nap at Coolham. *Photo: S. Woroniecki*

F/Lt Marciniak meeting Polish C-in-C General Sosnkowski at Coolham on 29 May 1944. Also present, S/Ldr Lapka (centre). *Photo: M. Cwynar*

Photo: B. Nowosielski-Slepowron

5

Operations from Coolham ALG, June 4–11 by 315 (Polish) Fighter Squadron

Flight Lieutenant Bozydar Jozef Andrzej Nowosielski-Slepowron has written this personal account using his pilot's log book, letters and diary notes written at that time, together with details obtained from 315 Squadron's Operations Record Books (RAF Form 541).

4 June 1944: No flying today, lousy weather, but something must be cooking. On the roads around the Coolham Airfield heavy traffic with loaded lorries and tank transporters all going south or slipping into surrounding woodland and hedgerows, covering themselves with camouflage netting and foliage. At night, heavy traffic of bombers.

5 June: Little flying today, weather not so good. Rain and wind. In between the squalls Dziubek sends few youngsters up, for very local formation flying. Late in the evening special briefing – INVASION on tomorrow. At night the heavy bomber traffic continues and one can see also large formations of Dakotas both solo and with gliders. Also some formations of Albacores towing Hamilcars and Horsas.

6 June: The weather in the early morning not so good. Everyone very excited. We listen to the speeches about the invasion on the BBC piped through the Tannoy. Everyone impatient, everyone wants to fly today, but though Horbaczewski implores W/Cdr Skalski and he phones Operations, we don't go.

The mechanics go over the planes for the nth time. Nothing in the morning or early afternoon.

About four, Tannoy message, "Briefing at 17.00 hours". All flying personnel gather at the squadron office tent by 16.30 hours. At 16.45 Dziubek came out and eased the misery by posting on the blackboard the list of the pilots that are to fly.

At first, recriminations, especially when we found out that S/Ldr Wierkiewicz was to fly as a "guest". "Why do we have to take these HQ wallies with us?!"

Those selected, preen themselves, happy as larks. Those to be left behind, at first resentful, soon start talking that the 1st eleven will fly later when the need for good pilots is greater. That leads to a few friendly insults and scuffles.

The entire Wing is to go. Ramrod 976, an escort to a glider drop in the Eastern beach area. Horbaczewski will lead our 315 Squadron and the central four.

The other fours are to be led by F/Lt Marciniak and F/Lt Stefankiewicz.

Others selected for this operation were: F/O A. Czerwinski, F/Sgt B. Czerwinski, F/Sgt J. Jankowski, F/Sgt K. Kijak, F/O B. Nowosielski, F/Lt J. Polak, F/Sgt K. Siwek, F/O K. Wunsche and the "guest", S/Ldr M. Wierkiewicz.

We, 315, and 306, led by W/Cdr Skalski, will be in close escort while 129 will fly higher to give cover to the whole lot.

We took off at 20.20 hours and met the glider train over Kent. 306 flew above them on their port (left) side and we, on their starboard (right) side in a gaggle of fours – 129 about 5,000 feet above and sunwards.

As they say – "Operation went according to plan". Only one enemy aircraft was seen and that one left in a power dive very quickly.

The gliders detached and landed to the south west of Caen. We had some light flak classified as intensive but inaccurate.

The weather was clear, no cloud, with slight haze and visibility 15–20 miles.

On the way back we saw a small ship, about two miles from the shore. The ship exploded and sank, not due to our action but probably due to long range artillery, ours or their's we did not know.

Landed quite happy at 22.00 hours, one after another in squadron formation. Had some extra food laid on in the mess tent. The talk in the mess was very happy and boisterous.

7 June: 05.05 hours first take off of the whole Wing. Ramrod 980. All bombed up with 500 pounders. The reconnaissance over Argentan, Buel, Dreux and Sees area was quite fruitful.

The weather was poor with bad visibility and 10/10 clouds at 1,500 to 2,000 feet inland and lower over the sea with top of cloud of several layers at 9,000 feet.

We had a direct hit on the railway station at Argentan, several more on the railway track and on a motor column.

One Fw190 came nosing around and F/Lt Sworniowski had all the luck and shot him down, later confirmed "Destroyed".

After releasing the bombs the same motor column was strafed with flaming results. Squadron came back in twos and threes and the planes landed with gaps of several minutes between various elements, from 07.10 to 07.40 hours.

The planes were refuelled, rearmed and checked by the ground staff at once and prepared for another sortie.

09.05 hours next take off, again at maximum effort. Ramrod 980 continued. 35 aircraft of 315, 306 and 129, all bombed up, took off. Our squadron bombed Dreux marshalling yards and a goods train standing there.

Middle of that train blew up with a terrific bang and lots of smoke! Ten or so Me109's were sighted about 15 miles south of Rouen at 4,000 feet, chased and attacked.

F/Lt Marciniak got one confirmed and one probable, F/O M. Kirste one confirmed, F/Sgt T. Berka one confirmed and F/Sgt R. Idrian one confirmed.

Once again squadron landing dispersed from 10.50 to 11.20 hours. Refuelling, rearming and checking aircraft was carried out at once.

Next take off, two aircraft at 14.10 hours for "weather recce" over the battle area. F/Lt Stefankiewicz and W/O Tamowicz come back at 15.15 hours reporting on their way back that the weather is satisfactory with cloud cover between 3 and 8/10 with base at 3,000 and top at 6,000 feet. Visibility between 8 and 10 miles.

At 15.30 hours our squadron takes off with W/Cdr Skalski leading the middle four and the squadron, with F/Lt S. Sworniowski and F/Lt J. Polak the other two elements. Two fours bombed up and one as cover.

Sixteen bombs were dropped at railway and road intersection about two miles south of Sees. Result, four direct hits and several near misses.

After that we flew around the area looking for trouble, but no opposition in the air or ground seen. All squadron back in formation, landing one after the other at 17.50 hours.

Aircraft rearmed and refuelled straight away ready for take off to finish the 980 Ramrod Operation.

At 18.45 hours 12 Mustangs of our squadron take off. F/Lt J. Marciniak leading the squadron and the middle four and the other element respectively by F/Lt M. Cwynar and F/Lt H. Stefankiewicz.

Eight aircraft bombed up and the other four acting as escort, attack the double railway track at Ecoucher, four miles south of Argenten. Sixteen bombs dropped with two direct hits observed.

All back at 20.40 hours. On return F/Lt J. Marciniak posted to 306 Squadron as Acting Squadron Leader to replace S/Ldr Lapka, shot down during the daily operations.

F/Lt M. Cwynar re-posted within squadron to take over appointment vacated by S/Ldr Marciniak.

8 June: At 06.50 hours F/Lt M. Cwynar takes off with eight bombed up aircraft for an Armoured Reconnaissance over Falaise, Alencon and Vire area.

Target Tourouvre road and rail junction. NB: S/Ldr M. Wierkiewicz flies with the squadron.

Weather improving: 5/10 cloud at 6,000 feet with five miles visibility. All back and down at 09.00 hours.

At 10.15 another eight bombed up Mustangs take off – Ramrod 982. S/Ldr Horbaczewski leads. Sixteen bombs dropped at rail and road junction near Vire. Near misses observed. Weather improving due west but deteriorating to the east. Back at 12.15.

At 15.20 F/Lt M. Cwynar takes off with another three bombed up Mustangs on continuation of Ramrod 982 to Falaise, Alencon and Vire area. Just south of the battle line Cwynar's aircraft is hit by light flak and trailing glycol smoke he force lands about seven miles west of Caen after dropping bombs inside bomb line.

F/O B. Nowosielski takes over the lead and the four escort Cwynar over the line. He is picked up alive and well by our troops on Caen-Falaise road. The three fly back over the bombing line and successfully bomb at U 154.

Weather continues to deteriorate – 8/10 cloud at 2,500 feet and visibility down to four miles. Back at 17.00 hours.

9 June: No flying today as the weather very poor.

10 June: Eight bombed up squadron aircraft take off in two separate fours – Ramrod 986. The first four, led by F/Lt A. Sworniowski, take off at 06.45 and the second, that of F/Lt J. Polak, at 07.10.

Over the bomb line the plane of F/Lt Sworniowski is badly hit by flak, he bails out and is picked up by crews of unidentified tanks.

Tanks did not open fire when the remaining three Mustangs flew over them.

The weather quite good – 4/10 cloud at 3,000 feet and visibility 15 miles. All seven aircraft back by 09.15.

At 11.35 four bombed up Mustangs led by S/Ldr E. Horbaczewski take off to continue Ramrod 986. Three miles south of Caen 10 tanks and about 40 M/T observed and dive-bombed and strafed. Near misses observed though light flak, heavy and accurate.

Weather indifferent: 10/10 cloud at 5/6,000 feet with patches of cloud down to 1,000 feet. Improving west of Caen. Back by 13.45.

At 19.30 take off of 12 bombed up Mustangs led by S/Ldr E. Horbaczewski, the other fours led by F/Lt H. Stefankiewicz and F/O K. Wunsche. Continuation of Ramrod 986.

M/T column observed about 1.5 miles south of Villers and immediately bombed. Bursts on target, one lorry seen to explode followed by flames and black smoke.

At 20.30 another M/T column observed on Alencon–Mamer road and very successfully strafed.

Nine lorries were seen to explode and about 80/90 per cent of the convoy believed destroyed. Weather: 5/10 at 3,000 feet with visibility of 10 miles. All back by 22.00 hours.

Today, F/Lt Sworniowski reported missing – he was never seen again. F/O B. Smidowicz, F/Lt G. Smidt and F/O Haczkiewicz posted and reported to this squadron and F/Sgt J. Zalenski posted to 84 GSU.

11 June: Very little flying today. Two aircraft Air to Sea gun test. Four aircraft air tested. P/O G. Swistun posted to this squadron from 306.

F/Lt Bozydar Nowosielski (left) with S/Ldr Horbaczewski.
Photo: B. Nowosielski

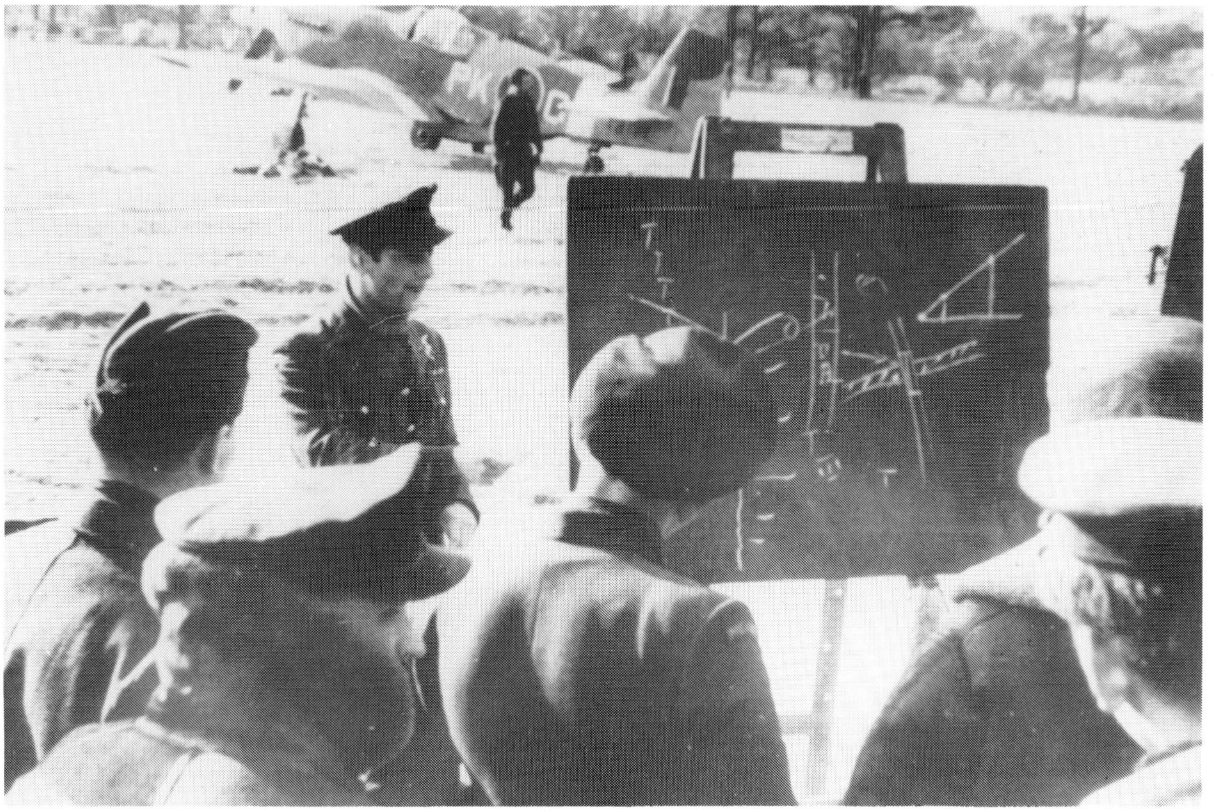

S/Ldr Horbaczewski briefing pilots of 315 Squadron prior to a mission. *Photo: M Cwynar*

D-Day, 6 June 1944, Coolham ALG. *Photo: B. Nowosielski*

Polish armourers of 133 Wing heaving bombs on trailers. *Photo: Mrs. M. Jarosz*

Photo: M. Cwynar

6

The "legendary" Squadron Leader Eugeniusz Horbaczewski

The Commanding Officer of 315 Squadron was Squadron Leader Eugeniusz Horbaczewski, P-0273, DSO, DFC, Silver Cross of Virtuti Militari and Cross of Valour with three bars (four times), Poland's fourth ranking ace of World War II.

He was an exceptional pilot and leader, highly respected by all who knew him.

His death when flying over France on 18 August 1944, only a few weeks after leaving Coolham, left everyone stunned.

His nickname was Dziubek, meaning beak, and also sometimes Horba.

One of those keen to pay tribute to him is F/Lt Douglas Coxhead, DFC, Ex 222 (Natal) RAF Squadron.

The Polish boys were all typical continental boys; correct, good manners and a wow with the girls.

We all met up at times, there were hundreds of squadrons around at that time.

Many lived from March to October under canvas on Sommerfeld tracking airfields. In the summer of 1943 the Poles flew Spitfires, my lot (182 Typhoons) were at New Romney, Kent.

One morning 182 had made a raid somewhere in northern France and had all returned.

Aircraft were being refuelled and pilots debriefed in a large marquee tent. Phones were laid on to group and many other squadron bases locally for liaison purposes.

The "Typhies" were apart – around 20 yards and facing each other about 15 yards apart, nose to nose – yet close enough to allow bowsers to refuel quickly.

Suddenly we were aware and very interested in a solitary Spitfire in our circuit.

We all piled outside and watched. A straggler coming in for a quick refuel? It often happened, then off to its base.

It taxied right up to our fuelling point and "The Pole" logged out.

I knew him, so did several others, as we'd had a couple of Poles fly with us "for a rest, for experience", earlier.

Horba bounded out, dress hat on at a rakish angle (I always flew with my hat in my pocket in case one landed away sometime).

He announced his arrival and refuel requirements to our C/O requesting it be done as soon as possible as it was an emergency. "One of my boys is down, in drink, in dinghy, many miles out. I have his position time from last beacon. I go back to him to give bearing and transmit to shore so that ASR (Air Sea Rescue) and Shagbats (Walrus flying boats) can get him before the Jerry does."

Radio transmission (R/T) was elementary and radar very much lacking then but a kite above a dinghy could transmit to bases ashore who would get a fix and initiate a rescue smartly. Otherwise "Jerry" could get a swarm of Focke-Wulf's and Me109's to attack him and/or the solitary aircraft circling above.

It wasn't a pleasant thought but it did happen at times.

His (Horba's) aircraft was refuelled smartly and in he got, started up and belted up to the runway's end, into wind and full throttle and took off.

He held it down so when the undercart came up the aircraft flew along at about 10 feet above the ground! It was just enough to keep the prop clear.

It was bloody good for speed attainment then up it shot to around 1,500 feet in a moment, did a "split arsed turn" downwind, circled our little airfield at 1,200/1,500 feet then dived down for his "beat up".

He put the aircraft at high speed between our row of Typhoons. The Typhoons were nose to nose at, say, 45 feet to 15 yards away, and he "buzzed up" the column of them (around 18 to 20 aircraft) at BELOW nose cone level.

The "Typhies" had a huge prop so when on deck, tail down attitude, the cone was about 12 feet above ground level.

It looked as though Horba would hit the ground but no, he flew his Spit up there over the ranks of aircraft BELOW prop height.

He "reefed it" into a near vertical climb at the end, did an upward chastie (slow roll going upwards), flopped over, straightened out and went off southwards to his "boy" in the dinghy alone and unescorted.

We heard later an ASR launch was almost due south of the Isle of Wight when a Shagbat got bearings first, landed and rescued the pilot.

Horba went back to his base. I never saw him again.

He was a fearless pilot, a gentleman with old continental good manners, almost clicked his heels, etc., yet a lethal machine who revelled in doing his thing for the pertinence of his war.

I was sorry to hear of his ending.

After 49 years it's as clear as yesterday.

HORBACZEWSKI THE POLE, I salute you!

❑❑❑❑

One of Horbaczewski's most flamboyant escapades while based at Coolham was his rescue of a fellow pilot who was shot down over France.

F/Lt Michal Cwynar recounts this incident, which happened in the middle of June, and also one other involving formation flying.

Our squadron, led by Horbaczewski, and also 306, led by Janusz Marciniak, were sent on a mission to strafe German positions south east of Cherbourg, to relieve pressure on the American positions.

While strafing German Panzer units our "A" Flight commander, F/Lt "Mysza" Henryk Stefankiewicz, was shot down by flak and killed.

W/O Tamowicz, his wingman, was also shot down and crash landed on a muddy field.

Our C/O, Horbaczewski, landed on what was a partly prepared landing strip nearby, helped Tamowicz out of his crashed aircraft, then loaded him onto the seat of his PKG Mustang. Sliding the seating compartment back as far as possible, Horbaczewski climbed in and sat on Tamowicz's knees, taxied back to get enough distance for take off and then was on his way back across the Channel to Coolham.

Back at our base, the 315 Squadron mechanics watched in amazement as Horbaczewski's "PK-G" taxied to the dispersal point with two pilots in the cockpit.

It was a great feat of courage and panache by our squadron commander. The name "Legendary Horbaczewski" will remain for posterity.

The "formation flying" incident occurred after the appointment of a new flight commander.

The replacement of flight commanders was usually decided by HQ but Horbaczewski would decide who he wanted then "suggest" the name of a candidate.

It certainly was the procedure when I was nominated to take over "B" Flight.

This time, our C/O, Dziubek, approached another dear colleague of mine, Henryk Pietrzak from 306 Squadron, to take over "A" Flight.

Early in the morning, two or three days after Henryk's appointment, Horbaczewski told his two new flight commanders that the three of us were going for some formation flying.

I knew what was coming because Dziubek had done this to me several times before.

From the Coolham village end of the runway, with Henryk on Horbaczewski's right and me on the left, we took off in close "V" formation.

Once off the ground, with our wing tips overlying those of our leader, we climbed to 8,000 feet in a southerly direction towards the Brighton area.

After fooling around for a while, Dziubek turned back towards our base.

Horbaczewski liked to show off but he had to have an audience. None of us was immune from this touch of vanity. We knew that down below, at Coolham, the critical eyes of 133 Wing fighter pilots were watching!

Diving gently, air speed just on the 400mph mark, Henryk and I as close as possible, Dziubek made one loop, diving lower and made a second. Pulling out of the dive at around 2,000 feet in a sweeping, left hand climbing turn, our leader decided to utter the only word during the whole flight "Ladujemy" – which meant we were going to land.

I moved slowly under the Mustang's fuselage, crossing over to echelon starboard, and with undercarriages and full flaps down and distances between each aircraft stretching, we flew over Coolham village, to land on our now so familiar Sommerfeld landing strip.

At the dispersal point, Horbaczewski got out of his aircraft, walked to his jeep and drove away.

Henryk Pietrzak, being a newcomer to our outfit, looked at me with a puzzled expression. I told him that it was our C/O's way of showing his approval.

Henryk lit his pipe and I, with the bad habit at that time, lit a cigarette!

One could compare that pleasant flying episode to the behaviour of three competent jazz musicians who felt like playing a "gig", a jazz session.

While playing, they responded to each other's feelings and moods. At the end of a session, there was nothing more to say. After all, they had just had a stimulating conversation.

❑❑❑❑

A fellow cadet with Horbaczewski at flying training school in Poland and who later met up with him at Coolham, was F/Lt Robert Stembrowicz, P-76732, of 315 (Polish) Squadron.

We were in different platoons while training and as the platoons led almost separate existences with different sleeping quarters, different hours of theoretical tuition and even different hours of flying, I am afraid I had not much chance of getting to know Dziubek, as was his nickname, well.

In my own judgment he was always a serious, quiet and very pleasant cadet officer. Not talkative yet not secretive, he was liked by all.

He was of "small stature", always standing at the far

end, on the left of a line when on parade, the tallest being on the right.

In England I did not come across him much (we were in different squadrons) until I was posted on 20 June 1944, from No. 25 EFTS at Hucknell, where I was instructor, to No. 315 Deblinski Squadron at Coolham ALG where Dziubek was a C/O.

In the short time he was my commanding officer I appreciated his calm yet imaginative leadership and his dedication to flying and care for his men, pilots and groundstaff alike.

Perhaps the best I can do to write about Horbaczewski will be to describe his last flight.

In August, No. 315 Polish Squadron was stationed at Brenzett ALG where it arrived on 10 July from RAF Ford. It was joined by 306 Polish Squadron and 129 (Mysore) Squadron. These squadrons formed 133 Wing.

From time to time these three Mustang squadrons operated individually over France, mostly on fighter sweeps "Ranger" combined with attacks on any ground targets.

On 18 August 1944, at 5.45am, the intelligence officer of 315 Squadron, F/O Tomszyk, woke up S/Ldr Horbaczewski with information that the squadron was to carry out a sweep over France in the region of Beauvais and attack any ground targets that presented themselves.

Dziubek had been ill for some days with flu and on that day he looked terrible. His face was grey, eyes sunk deep in his face, and he was breathing heavily and with difficulty. The squadron medical officer had forbidden Dziubek to fly some days previously and all who saw him that morning tried to stop him. Indeed, F/O Tomszyk begged him not to fly.

Dziubek could be very stubborn when he felt like it and all our pleading was in vain.

"I am going," he said, looking like death warmed up.

Collecting his maps and flying helmet Dziubek quietly whispered to Tomszyk: "I tell you, if we meet Germans I shall not come back, my legs are giving way and every time I lower my head I black out."

Tomszyk used some strong words, not the words to use to one's squadron commander, but to no effect. "I am going," was the short answer.

At 07.20 hours 12 Mustangs, call sign "Central", took off from Brenzett on Rodeo mission 385. In 15 minutes they crossed the "exit gate" over Brighton and set course for France.

At 07.47 hours, at 8,000 feet, the squadron crossed the French coast over Le Touquet. Sweeping in a wide arc in the region of Beauvais, W/O Slon reported enemy aircraft taking off below (he noticed the waves of grass made by aircraft).

F/Lt Pietrzak, the C/O's deputy, heard the warning but Dziubek did not react. He assumed that the C/O's radio was not functioning. He ordered W/O Slon to attack and followed with his four ordering the second and third four to cover them from above.

As they saw the number of Fw190's, some taking off, some already assembling in the air, the whole squadron attacked.

Mustangs attacking from the light and with the element of surprise, created havoc.

Dziubek's No. 2 saw him shoot down three enemy aircraft but in the fight and himself attacked, soon lost sight of his C/O.

F/Lt Pietrzak called on the radio for the squadron to reassemble over Beauvais at 8,000 feet. In due course single P-51 Mustangs joined him. That is, all but one – the C/O was missing.

On returning to Brenzett the details of the flight became clear. The flight lasted 1 hour 55 minutes and the result was 16 Fw190's destroyed, 1 Fw190 probably destroyed and 3 Fw190's damaged.

Individually this added up to:

S/Ldr Horbaczewski	3 destroyed
F/Lt Pietrzak	2½
F/O Nowosielski	1
W/O Slon	1½
P/O Swistun	1+ 1 probable
F/Sgt Kijak	1+ 1 damaged
F/Lt Schmidt	1
F/Sgt Bargielowski	2+ 2 damaged
F/Sgt Siwek	3

There was no news about Dziubek until some years after the war.

Later information indicated that he was shot down over Velennes (Oise).

The aircraft was buried deep in the ground. The Americans, cooperating with the French Red Cross, excavated the plane.

The remains of the pilot were found below the plane, which seems to indicate the plane hit the ground in an inverted position.

S/Ldr Horbaczewski was buried in the military section of the town cemetery at Creil. The number of the grave was Z 379. Capt Pedder witnessed the exhumation, and from the RAF side, Sgt Whitehead, No. 1 MREU.

The fight over Beauvais was the biggest victory by any squadron in one flight during World War II.

Horbaczewski was honoured – even by the Polish Communist government – who named a street after him in Wroclaw and Warsaw and a Liceum (Higher School) of Air Force in Zielona Góra (Green Mountain).

Dziubek had on his "account" 16 aircraft destroyed, 1 probably destroyed and 1 damaged.

This ranked fourth among all Polish pilots' victories during the whole of World War II.

According to British statistics one Polish fighter pilot lost equalled three and two-thirds German planes destroyed and 10 aircrew killed.

S/Ldr Eugeniusz Horbaczewski, DSO, DFC.
Photo: J Bargielowski

High stakes bridge at Coolham – S/Ldr Horbaczewski enjoying a game with groundcrew.
Photo: P Skrzypczak

7

306 Torunski (Polish) Fighter Squadron

Squadrons landing at Coolham for the first time landed in style.

Squadron Leader Zygmunt Jelinski, of 306 Squadron, tells of 18 Spitfire aircraft leaving Heston RAF Station bound for Coolham on 1 April 1944. He recalls the horrific death of one of the pilots who could not escape his blazing 'plane but also, on the lighter side, several amusing incidents including when his own Mustang appeared to land backwards.

Flying in "parade formation" we made our final pass over Heston and said goodbye to the comforts of home. We set course south and about half an hour later were looking over our new base which, from above, without the usual buildings and facilities, looked a little deserted.

Nevertheless, the surrounding country was beautiful.

Over Coolham, in preparation for landing, we dispersed and aligned astern to make our respective landings on a steel net runway.

Our advance party had already erected some tents for personal use, dispersal points, maintenance and other facilities. We would be living under canvas and be fed from a field kitchen.

A few Mustang III's had been delivered and the process of retraining was about to begin.

The Spitfires were ferried out to other locations by pilots from the Air Transport Auxiliaries, the ATA being a non-military organisation staffed by all kinds of assorted pilots including women. They had to be very versatile and they did a great job, delivering a Spitfire one day, then flying a four-engined Lancaster the next day.

At Coolham we saw one of our Spitfires taking off and immediately knew the pilot had forgotten to change the propeller pitch from "coarse" to "fine" resulting in loss of power.

He just managed to leave the ground, but unable to clear the incline he crashed beyond the boundary of the airfield.

His undercarriage was already up so we knew that the crash was not serious.

Some of the rescuers then told us that the pilot was not hurt though he had not left his cockpit.

When they reached him – he was an Englishman – all they could hear was a series of assorted swearwords emitted at a rate of 1,000 words a minute.

The only damage was to his artificial leg, which was broken.

His violent reaction was due to the fact he had just received his new leg, after waiting six months for it and only the day before he had begun flying again. No wonder he was upset!

I was a senior pilot in "A" Flight, commanded by my very good friend, F/Lt Zygmunt Drybanski, a veteran of many air battles.

He also had a short temper, as on one occasion when, not being satisfied with the squadron's performance in precision formation flying, as punishment, he led the 12 aircraft in a series of group aerobatics, including a 12 plane loop. It was quite a circus!

On 3 April he asked me to be in charge of four other pilots and to bring five Mustangs back from Aston Down to Coolham. We were to go by train to Swindon, stay the night in the Station Hotel, and fly back to Coolham the next morning.

My fellow pilots, all contemporaries from the Polish Air Force College, were Flying Officers Andrzej Beyer, Stefan Tronczynski, Wladyslaw Szajda, from my own squadron, and Adam Sworniowski from 315.

Andrzej Beyer had an air of authority about him. As the son of a landowner in Poland he was used to giving orders, though on this assignment he was my subordinate.

Stefan Tronczynski had a bold, challenging stare, was arrogant, strong, always spoiling for a fight. He took part in many a bar brawl, usually over women. His weakness was air sickness.

Wladyslaw Szajda was serious about himself, rather a loner but an excellent pilot, very tense and determined to succeed. I felt that for some reason he resented my authority, thinking perhaps that it should be his, and yet he often flew as my wingman and did an excellent job.

Adam Sworniowski was rather fastidious about himself and his dress. He was well bred with impeccable manners but unfortunately his voice had a Donald Duck quality, which somewhat spoiled the debonair appearance.

I got along very well with all of them, but then I am a born diplomat. We passed the time playing cards and talking about our exploits though Szajda and I seldom interrupted the flow of conversation.

After registering in the hotel we discovered the town was swarming with Americans. GI's were everywhere, leaning against every available wall in town.

Szajda and Beyer decided to retire early but Tronczynski, Sworniowski and I were dying for a beer.

Unfortunately, the lounge of the hotel was full of American officers and when we sat down at a free table we learned that the only drinks served in the lounge were martinis at four shillings a shot, which, compared to a shilling a pint, was very expensive.

Well, we were not about to drink that stuff at that price, so we left.

But luck was with us and about 9 o'clock when I was alone in the hotel's writing room an elderly butler approached me and asked if I would like a pint of beer.

He took me completely by surprise but I answered that I would be delighted and he brought me a beer that tasted the best of any beer that I have drunk since.

The butler also agreed to serve my fellow officers but only Tronczynski and Sworniowski came down to join me and when "Time" was called in the neighbouring bar more hospitality was extended to us by the landlady who invited us to be her guests in a room behind the bar.

There, in the company of her two beautiful daughters (at least, after a few drinks, they seemed beautiful), we continued the party.

That night we felt that we had been embraced by the cold and reserved Britons as part of their own family.

Those were very touching and warm moments, especially that the landlady offered us a bottle of Polish vodka which she obviously had been saving for her business.

Knowing that we were flying the next morning we could not possibly drink it but she still insisted that we take it with us. The party must have broken up about one in the morning but the warm feeling has stayed with me for all these years.

The next morning I took off for Coolham with Sworniowski and Tronczynski forming the first section and Szajda and Beyer were to follow.

After circling the airfield I decided to set course for Coolham, hoping that the other two would catch up with us. When we landed at Coolham there was still no sign of the other two.

About an hour later one of the other Mustangs, piloted by Beyer, landed.

Visibly shaken he approached me and immediately began the gruesome story of Szajda's death.

Beyer was already airborne when he saw that Szajda's plane crashed on take off, beyond the boundary of the runway.

He immediately assumed that perhaps Szajda did not switch over to the main petrol tanks, and his engine cut, although Szajda was too good a pilot to make a mistake like that.

He landed at once and ran over to the disabled plane. Szajda was still in the cockpit, his feet trapped by the engine.

His auxiliary tanks had broken off on contact and caught fire some distance away from the aircraft but the flames slowly began to creep towards the disabled plane and the trapped pilot.

The ambulance and the fire engines were already there, but the fire brigade were unable to contain the burning tanks and by some freak chance the fire engine itself caught fire and started to burn. Another fire engine had to be brought in.

In the meantime F/Lt Carrin, the RAF Station's medical officer, was making a gallant effort to free Szajda, who was in terrible pain, from the wreckage.

Finally, he began to amputate his feet, but the heat from the burning aircraft proved to be too much and he himself fainted inside the cockpit.

He was pulled out from the burning wreckage and revived. He then injected Szajda with a strong dose of morphia, who then went to his burning death.

As much as we were used to our pilots being killed in action, we were all sick listening to Beyer's report. Nobody deserved less to die in such suffering as Szajda did.

Apparently a Court of Inquiry was held at Aston Down, attended by F/L Lt Sporny and F/Sgt Michalkiewicz but I don't remember what the final verdict was.

(Extract from RAF Form 1180, Accident Record Card states that an A.I.B. investigation concluded that the pilot took off with an engine which was probably insufficiently warmed up, thus resulting in total power failure.)

Training and testing on the Mustangs, fitted with bombs, continued for all squadrons and it did have its lighter side.

We were flying in two's and one of the pilots from 129 Squadron got in the vortex of the propeller, lost control and crashed into the latrine, unfortunately, but everything was OK.

A couple of hours later one of our pilots did exactly the same thing but turned to the left earlier and the engine made a terrific racket and finally seized.

We ran over to rescue the pilot and couldn't find him. Finally we found him hidden behind a tree.

I asked him what happened and he said he was waiting for the bombs to explode.

I said, "Look behind you", and there they were, two of the bombs, lying there. Of course, he nearly had a fit!

Then one day one of the mechanics asked me if I could test fly a plane. I said sure, it was only a five-minute flight. What I didn't realise was that it had long range tanks suspended.

I took off for the flight and I came back using about the same approach speed as normal.

But the plane was very heavy and when I dropped it down it jumped back about 20 feet in the air. I gave it more gas and came round again and then I saw in front of me that latrine.

I said: "Oh, my God, I'm not going to kill myself in a situation like this." So I shut off the engine and I put the right brake on. I was off the runway by this time. The plane circled and faced the take off direction but was right back on the runway and then it started to move tail first so it looked like I just landed the Mustang tail first.

As it turned out there were some witnesses. One of them was an Air Commodore and he said to the Group Captain: "What is this, have you seen this?" And the Group Captain said: "Oh, don't take any notice. It's only one of the bloody Poles, showing off as usual."

Of course, I was the laughing stock of my squadron for a while, landing a Mustang backwards, but I never damaged it.

As well as training, additional flying duties were being outlined to the pilots – long range escort duties with flights up to six hours' duration and dive-bombing missions where instead of two 90 gallon tanks we would have two 500lb bombs suspended under the fuselage.

To increase combat accuracy a new gyroscopic gunsight was introduced and I was sent on a short gunsight instructor's course at the RAF station in Southend so I could instruct other pilots when I returned.

At Southend I was flying a Spitfire Mk IXb as at the same time they were also testing the 120 octane petrol which contributed more power.

I did get myself into a little bit of trouble with the station's Commanding Officer who had me on the carpet for doing low level rolls off the deck.

With all that power and a Spitfire that was not overloaded with armament I could not resist from showing off a bit.

Back at Coolham further intensive training continued during which we pushed the new aircraft to its limits. As a result there were more accidents.

One day, Wladek Potocki, who had just taken over the command of "A" Flight from Zygmunt Drybanski, came to me and Karol Marschall and said: "Let's see how high the Mustang can fly."

Now, Wladek Potocki was a brilliant man. He was not only the top student at the Polish Air Force College in Deblin, an exceptional pilot (later to finish as the No. 1 graduate of the Empire Test Flying School, test pilot in the Farnborough Air Shows, and for five years the chief experimental test pilot with Avro Canada), but also a great leader.

Even his appearance was striking. He was quite tall with broad shoulders, jet black curly hair and characteristically hooked nose.

By contrast, Karol Marschall, a few inches taller, was quite slim, with fair hair, sallow complexion and a very quiet demeanour.

The three of us, old friends, took off in P-51c Mustangs and started a long climb.

We reached 40,000 feet and we were unable to climb any higher. At that altitude our Mustangs were extremely sluggish and each aircraft behaved as though it would stall any moment. So we began a shallow dive to about 27,000 feet, still flying in "V" formation, when Wladek began looping the loop.

Here, my recollections differ from Karol's, as he claims that we were at 35,000 feet when Wladek pulled up and started a barrel roll, rolling towards him.

I think that my version is probably the more likely one as Wladek was too smart a pilot to try a barrell roll in such rarified air, especially as we had never done that in a "V" formation, whereas we had done the "V" formation loops before.

Anyway, when we were upside down at about 35,000 feet, each of our aircraft had stalled and immediately went into a spin.

Wladek and I recovered almost immediately, but Karol was nowhere in sight. Only when we had landed in Coolham, and waited for his arrival, did we find out that he did not recover from an inverted spin until he reached about 8,000 feet, which was a hell of a drop!

He was very shaken and pale when he got out of his Mustang and walked over to our dispersal tent. I don't think that he expected to come out of this alive. He remembers it vividly to this very day. No wonder!

When there was no flying Stefan Tronczynski, Zygmunt Drybanski and I used to go to town, usually starting our rounds in our favourite pub, the Black Horse, in Horsham.

I remember the very first visit there, where in the saloon bar we found ourselves quite ostracized by the local gentry.

Fortunately, that did not last too long as quite a few of the local beauties found out that both Stefan and Zygmunt Drybanski were very entertaining.

Another diversion for some of us was bridge, organised by the padre, S/Ldr Stolarek.

Andrzej Beyer, Jurek Polak (315 Squadron) and I, enticed by a bottle of sacramental wine, became his regular partners.

In May and June it stayed light quite late, so he set up a card table in front of his tent and the four of us, sipping the sacramental wine, usually played about three rubbers.

After one such evening Beyer had returned to his tent, which he shared with Henryk Pietrzak, lay down with a cigarette in his mouth and promptly fell asleep.

An hour or so later we were all woken by blood curdling screams, which were followed by a ghostly figure leaving the tent and proceeding at a full run to the sick quarters, located some three quarters of a mile from our camp.

It was Beyer, who suffered superficial but very painful burns caused by his smouldering canvas bed. It was a miracle that neither Pietrzak nor he had suffocated for the tent was filled with smoke.

Anyway, next day we all had a good laugh over his misfortune.

Not all such incidents had a happy ending.

By the middle of April we were almost fully operational again and every day were at "readiness" – 30 or 60 minutes. In between, we practiced squadron formation flying and air-to-air firing.

On 19 and 21 April some of the pilots took part in combined exercises with the army which included air-to-ground firing, strafing attacks on ground targets and simulated dive-bombing.

Although we all loved the Spitfire, because it was the most beautiful aircraft ever built, when we began to push it to the limits of its performance we had found out that if, in a dive, the speed of 450mph was exceeded, the ailerons would begin to bend up and one was in danger of losing one's wings.

In order to provide some kind of an early warning the inner edge of an aileron was painted red and when the pilot saw red he'd better pull out of a dive at once!

Therefore, we were very happy to learn that with the Mustangs such restriction no longer applied and we could dive at speeds up to about 480mph without any fear of structural failure, or so we thought.

P/O Brygider joined the 315 Squadron on 7 April and on 19 April he was a member of a section of four aircraft practicing battle formations.

In a power dive his Mustang experienced a structural failure of the vertical stabilizer. P/O Brygider could not recover from the power dive and was killed instantaneously.

Immediately we were ordered not to exceed 550mph in a dive. The stabilizer was redesigned and suitably reinforced a few months later.

Another fatal accident occurred on 23 April involving F/Sgt Osieleniec, a newly arrived pilot with 306 Squadron.

He was coming in to land on the NE-SW runway, after completing a training flight, when he struck a tree stump with his right wheel.

When the airfield was built, a number of trees had to be cut down and the approach was over the residual tree stumps.

Since the runway was short we all tried to touch down right on the edge. Osieleniec tried to do the same thing. Unfortunately the wheel, after the impact, struck and damaged the right flap. Not knowing the extent of the damage he immediately gave full throttle electing to follow the standard procedure of going round again.

From our dispersal point we helplessly watched as, with the increasing speed, the undamaged flap began to act as an aileron, forcing the plane into a steep right turn and the pilot to struggle trying to keep it straight and level.

Finally, at an altitude of about 500 feet, his strength was overcome by the superior dynamic force and, to our horror, his Mustang, in a steep power dive, hit the ground just beyond the airfield's boundary. A column of black smoke, marking the spot, rose to an altitude of 1,000 feet.

The first operational flight from Coolham was made by F/O Jerzy Polak (my friend since before the war after meeting on a glider pilots' course) and F/Sgt Tadeusz Slon (later promoted to Warrant Officer), both from 315 Squadron, who were scrambled on 2 April, still flying Spitfires Vb, to investigate enemy activities over the Channel. They returned after 55 minutes without encountering any opposition.

On 26 April the distinction of the first operational flight from Coolham on P-51 Mustang aircraft was again with the 315 Squadron.

This time it was F/O Michal Cwynar who led a section of four aircraft to take part in Operation Ranger, the objective being to destroy enemy aircraft in the air or on the ground in the Albert–Cambria area.

F/O Sworniowski, F/Sgts Berka and Siwek, were the remaining pilots under his command. Jurek Polak and W/O Adamiak patrolled the English Channel as the Air Sea Rescue section.

During the two-hour flight Michal and his section did not encounter any opposition in the air but in spite of heavy flak they attacked some ground targets and then all returned safely to base.

Michal was a very experienced pilot and an excellent leader respected by both his subordinates and his superiors.

Zygmunt Jelinski, F/Lt P-1644
California, USA.

❒❒❒❒

Flight Lieutenant Lech Laszkiewicz has vivid memories of his time at Coolham.

In early Spring 1944, I was a Pilot Officer, based with 306 Squadron at RAF Heston, a few miles to the west of London, flying Spitfires.

There was a lull in air activities and our operational sorties were about equally divided between enemy territory in daytime and the Windmill Theatre in Shaftesbury Avenue in the evenings.

Vague rumours started circulating about an imminent change of aircraft, from Spitfires to the North American P-51 Mustang III's.

In due course the rumours became a reality. The first Mustang arrived together with one single, solitary copy of its technical manual, to be shared by the three resident squadrons (129, 306 and 315). This book soon became a treasure more precious than an illicit copy of the Esquire in a boarding school dormitory.

As much as we loved our trusty old Spits, the new Mustang was a very impressive aircraft. It had an enormous range and endurance and many built-in comforts. The "Brass", of course, had the first shot at the book of words and gradually the wisdom was eked out to us, the lesser fry.

We started to swot diligently, learning what and when to press, or not to press, where to put the fuel, oil, coolant, how to make it go up and which page to turn to for instructions to get it down in one piece and preferably unbent.

While the familiarisation was going on, orders came to move our three squadrons to a newly constructed Advanced Landing Ground at Coolham, near Horsham in Sussex.

We ceased to belong to Fighter Command and became 133 Airfield, 2nd Tactical Air Force.

Our lifestyle changed overnight, from the home comforts of permanent RAF stations with their many amenities and comfortable messes, to Boy Scout like existence among the green fields of Sussex.

True, we were not issued with short trousers nor woggles but we slept in canvas tents, in canvas camp beds, we washed and shaved in cold water in canvas buckets, we ate in a canvas tent mess and performed the daily necessities of life in hedges or canvas screened Elsan bucket latrines, sited near the westerly end of the wire mesh covered grass runway.

Occasionally the more hardy souls who dared to, took a bath in a miniscule collapsible canvas bath, al fresco, in front of the mess and sleeping tents.

This last activity, on one occasion, brought an attack of maidenly blushes on the face of a comely young female ATA ferry pilot. Having just delivered an aircraft she was being led to our mess tent for lunch. The girl was covered in confusion and the bather in a few suds, the bath being too shallow for deep diving.

The field latrine also provided some hilarity. One intrepid aviator was availing himself of the facility while a Mustang at the other end of the runway was run up to full power prior to take off. Suddenly, its heavy four bladed propeller broke free from the engine. The energy of some 1600hp stored in the rotating prop caused it to describe an elegant, shallow parabola in the westerly direction, terminating on the ground among the Elsan buckets of the latrine.

The "Loo-tenant" was unhurt but somewhat discomfited and heavily scented. We all walked up wind from the poor man and no one would shake his hand for weeks.

On the plus side, following the immutable laws of physics, the considerable residual energy remaining in the spinning propeller was transferred to the Elsan buckets and their contents. A large area of Sussex was thus liberally sprinkled with high grade fertilizer, a free and unsolicited manna from heaven, much appreciated by the farming community.

Joking apart, we got down to serious work to make 133 Airfield an efficient operational unit. The three squadrons integrated happily, the 129 Squadron personnel managed to survive the culture shock of eating Polish dishes as concocted by our "Chefs" and even acquired a taste for some of the more exotic delicacies.

New Mustangs were being ferried in from Aston Down by those of our pilots who had already accumulated enough experience. Conversion training proceeded apace and in spite of a few accidents and some fatalities we were soon ready for action.

Before we could fly on operations there was one problem to be solved. The oxygen system and oxygen masks combined with microphones, as used on Spitfires, were not compatible with the Mustang system.

American masks and microphones were not compatible with the English radio installation as fitted to our Mustangs. A quick "fix" modification was devised to the aircraft system and to our oxygen mask and I was "volunteered" to take the modified Mustang as high as it would go to test the new arrangement.

I don't remember how long it took but eventually the unpressurised Mustang staggered to 43,200 feet. The day was very clear and the view was incredible. I could see the whole of England and Ireland from coast to coast not to mention the continent. The sky turned very dark blue, almost like the night. The Mustang was wallowing, hanging on the prop, and didn't want to go any higher.

I was feeling a bit woozy, "pins and needles" were affecting my joints. It was time to go down. The last part of the test was to fire the guns at high altitude. I squeezed the trigger on the control stick and the six half-inch calibre guns rattled away.

Well! The guns were OK but the recoil force of the combined artillery (each gun firing 600 heavy bullets a minute at very high muzzle velocity) robbed the aircraft of its already minimum flying speed.

The Mustang promptly fell out of the skies, out of control. Some moments of fright and 12,000 feet later, things sorted themselves out and the dropping brick became a flying machine again.

Back at Coolham the modification was judged a success. 133 Airfield was ready for operations in earnest and I had learned that firing the guns was one good way to slow the sleek Mustang in a hurry. This knowledge proved to be very useful to me in the near future.

Our Spitfires were flown away and we proceeded with operations on Mustangs. The "steeds" proved to be very fast and comfortable on long flights, the fire power was impressive but most importantly the freedom from worry about fuel reserves was a great blessing. With all tanks full and the external under-wings long range tanks fitted, one was likely to die from starvation before the fuel ran out.

Perhaps for that reason, in case of need, a "pilot's relief tube" with a neat little funnel was fitted in the cockpit. Some young visitors from nearby Christ's Hospital school were seen to examine in detail what they thought was a "speaking tube". Nobody wanted to enlighten them!

As time went by, the character of operations changed and we started to carry two 500lb bombs under the wings, looking for likely targets. On one particular flight with bombs, low cloud prevented us from bombing and we returned to land at Ford where the Royal Navy personnel were to off-load the bombs while we were having breakfast in the ward room.

Next thing, panic started on the field and we were given some very dark, unfriendly looks. It transpired that our aircraft were armed with non-standard pressure bombs which were meant to explode not on impact but a few feet off the ground, triggered by the air pressure wave formed between the surface and a sensor on the nose of the bomb.

The bombs were not to be landed with under any circumstances. Scrambled eggs were not only on our breakfast plates but on our faces as well.

We were duly directed to the nearest air raid shelter while the Royal Navy sweated and no doubt expressed their views on our parentage. Luckily, all ended without a mishap and we returned to Coolham. I don't remember landing at Ford again!

One morning I was woken in my tent by a sharp sting on my middle finger. As I opened my eyes I saw a spider scuttling away into the grass under my cot. By next day my finger was the size of a salami sausage and about the same colour. A visit to the doctor's tent produced some "Hms" and "Ahs" followed by a disinfectant swab and bandages. Antiobiotics were still a thing of the future. By the evening my whole arm was throbbing and the finger was getting bigger and more painful. I was off flying for quite a long time. I nearly lost my finger which is still bearing scars and I HATE spiders to this day.

By now we were flying over France, Belgium, Holland and Germany, sometimes just to provoke the Luftwaffe fighters to come up and meet us, sometimes bombing railway yards and important junctions, sometimes escorting our bombers on the way to Germany and back.

While off duty we enjoyed forays to Brighton and sight seeing in the country. Gradually it became evident that things were coming to the boil. Road verges, centre reservations on dual carriageway roads and any available stretches of grass were filled with thousands of tanks, armoured cars, amphibious vehicles, lorries, jeeps, artillery pieces, ambulances, field kitchens and all manner of war hardware.

On the morning of 5 June we found 133 Airfield full of unusual activities. Our aircraft were hastily painted with wide black and white stripes on the wings and fuselages. We were all confined to camp and external telephone lines went dead. We guessed that the balloon was about to go up, the long and eagerly awaited invasion was about to commence.

We sat on the ground, fidgeting, waiting for things to happen. Evening came and nothing transpired. In the end, somewhat disappointed, we bedded down in our tents for the night. Sleep did not come easily and in our excitement we were dozing fitfully.

Then we heard an ever swelling drone of heavy aircraft passing overhead towards the coast, wave after wave.

This was IT! Operation Overlord had begun. To us Poles, it was an event of unparelled joy. At last we were about to begin retracing our steps back home to Poland. The rest of the night was passed talking and smoking innumerable cigarettes. Morning came and everybody was anxious to hear the latest news from France. Once again we sat waiting and guessing all day.

At 20.00 hours S/Ldr Lapka led 306 Squadron to France to form air cover for the landing invasion forces. 315 and 129 Squadrons were also taking part in this operation. They returned to Coolham some two hours later. Next morning, 129, 306 and 315 Squadrons were airborne at the crack of dawn armed with two 500lb bombs for each aircraft.

Two hours and 15 minutes later the squadrons returned to base, having successfully completed the first part of Ramrod 980.

The squadrons were airborne again at 09.55 to carry out the second part of Ramrod 980. Once again we carried bombs. The weather was overcast with occasional light rain. We were flying low, in a wide, almost line abreast formation.

Having crossed the French coast at "0" feet we proceeded southwards, inland, towards our target. The cloud base was low, about 600 feet. Shortly after crossing the coast we observed a number of Me109's flying in tight formation, just under the cloud base, from the west to east.

They did not seem to have seen us and were not taking any action. This appeared to be a golden opportunity for a quick score. As each of our aircraft in line abreast turned left to follow the 109's, we became a line astern, following each other at full throttle, still very low to avoid being seen.

Going into a scrap burdened by two 500lb bombs was not considered good judgement. Naturally, everyone jettisoned their bombs. (The bombs can be dropped un-fused, jettisoned, and, with a bit of luck, they don't explode on impact, or they can be dropped live, in which case they can be nasty and bite.)

The bombs dropped from an aircraft in front of me turned out to be nasty and hit me. One or both of them exploded right under my Mustang. There was a terrific bang. A "gigantic" sledge-hammer blow turned the aircraft upside down. Flying at 350 mph, upside down, at less than 50 feet with the windscreen covered in mud and no windscreen wiper, was not my idea of fun nor fair play.

I felt a little miffed, to say the least. I rolled the Mustang on an even keel. When my eyeballs stopped bouncing, a quick scan of instruments revealed everything apparently in order, the engine was running, the wings were still in place and unbent.

Ahead, the Me109's were still there, not taking any notice. All's well that ends well, I thought.

A glance in the rear view mirror revealed nothing untoward except for some debris still descending from the skies. So I kept going after the 109's. When I was just about to pull up to line up for a shot at the nearest Messerschmitt, my engine coughed and the oil pressure needle slowly descended to "0". Not my day, I thought, it's time to hightail home to safety, to be able to come back and fight again. A smart turn left to the north.

Some quick and important decisions had to be made. The engine without oil won't run more than a few minutes, if nursed. Won't make it to England, for sure. Where is the nearest beachhead?

Quick "Mayday" call on the emergency frequency on the radio, with request for homing to the beachhead.

A sweet voice of a WAAF says: "Return to base." Some hope! Climb into cloud and bale out when engine stops? Not a good idea. Anything coming out of the clouds on a parachute during invasion must be a paratrooper and a legitimate target for a turkey shoot.

It seemed wiser to keep low, at least to the French coast and to nurse the engine, hoping it would last a while longer. Once over the sea would be a good time to climb to a safe, parachuting height and to keep my fingers crossed.

I pulled the rpm lever back as far as it was practicable and settled down to see what fate had in store for me. The cloud was still low, about 300 to 400 feet. I went hedge hopping, watching the rear view mirror, for that's where the danger lay.

Being on my own I felt very vulnerable. I needn't have bothered. The Me109G came head on, also at "0" feet. Oh boy! I thought, it never rains but p....s, sorry, pours!

His cannon and guns started winking at me as I squeezed my own trigger. With the combined closing speed of some 700mph it only took a fleeting moment before we passed each other with no discernible harm done on either side.

Being in no state to tangle with the guy I kept going for the coast, thinking, "If he leaves me alone, I will leave him in peace." Looking over my tail in the mirror I saw the Me109 pull up to the cloud base and begin to roll over. He had other plans for me!

There was no choice now. I pushed the rpm lever and the throttle to full power and banked, pulling the stick to my navel in a steep turn to meet my adversary head on again. Meanwhile, he rolled upside down and pulled his stick tight to execute a roll off the top thus reversing his course to get on my tail.

He was now pulling hard out of a vertical dive at less than 300 feet, trying to get his twinkling cannon to bear on me, while I was rolling out of a vertically banked turn to face him and squirting my six .5 guns at him. Well, he almost managed to pull out of his dive but not quite.

The 109 "greased" itself into the ground, still firing and disintegrated in a ball of flames. Poor guy, I didn't think! You shouldn't have joined.

I was now facing south, away from the coast and relative safety. I completed the turn to the north and pulled the power back again. All this takes a long time to narrate but it all happened in seconds.

Hope springs eternal, the good Packard Merlin engine was still running and the coast could not be far! Just as I was beginning to relax a little, the engine roared to a crescendo, out of control, revs way past any red markings and book limitions. (Rpm of the Merlin are hydraulically controlled by the engine oil contained in a regulator in the propeller hub.)

There was no nursing the engine now. Under the circumstances, the maximum distance could only be achieved at full power. So, there I was, at "0" feet, roaring at "full bore" at 400mph. The French coast appeared in the distance. I was approaching the coast at a right angle, my line of flight was taking me towards a narrowish peninsula, with some cliffs at its end dropping steeply to the sea. Good, I thought, only a few more miles. Just then the thoroughly abused, overboosted, over revved engine gave an almighty shudder and screeched to a dead stop.

I had enough speed to pull up into the clouds, to a height sufficient to bale out, but had already decided not to be a "sitting duck" target for the heavy and well alerted coastal defences.

We were told that the Mustang, with its large cooling air scoop under the belly, is not the best bet for ditching. Belly landing was the only alternative. Easier said than done.

There was a line of tall high tension electricity pylons between me and the coastal cliffs. To pass over them would leave insufficient distance to come down gradually and to slow down for an "arrival". There was a good chance of nose diving over the cliffs into the briny.

I didn't fancy shredding and frying myself going through the 400,000 volts, steel cored cables. There was only one way, go under the thick cables through some thinner wires, becoming visible as I approached.

Oh, dear! The fields in front and beyond the cables look like a porcupine's backside, sown with hundreds of hefty anti-invasion telegraph poles. Just like the ones used around the coast of England for the same purpose.

Nothing for it now! Speed, 280mph; fuel cocks, off; ignition, off; wheels, up; flaps, won't work, no power and much too fast. Speed, 250mph; jettison the hood. Too fast, too near the wires. SLOW DOWN! Fire all guns, that will help and is also likely to keep down the Wermacht sharp shooters. Speed 220mph, guns thumping, 200mph, aim into a gap between the telegraph poles. Nose down into the field and head down. Banging, thumping, guns roaring, screech of torn metal, splintering, crashing telegraph poles! GOODBYE!! Silence . . . .

A tweet, tweet, tweet of a bird, and humming of the wind in the wires. Sound of sea waves breaking on the cliffs!

Hey, I am still here! Bruised but apparently in one piece! Head up, look out of the cockpit. Not so good! What is left of the torn off wings is straddling a German dug-out trench. Fuselage all wrapped up in coils of barbed wire with a notice: "Achtung Minen!" plus skull and crossbones entwined in them. Crackling of the cooling hot metal.

Get out and burn the aircraft, as you were told.

Too late!

Grey uniformed figures climb out from the dug-out, one points a sub-machinegun at me and says: "Cum, ze var she is over for you!"

Eleven months later I am cautiously creeping through the Russian front lines, on my way from the recently "liberated by the Soviet Army" Stalag-luft One, towards the British forces in Germany.

I didn't fancy a free ride from our POW camp on the Baltic coast, via Poland, to a gulag in Siberia being offered by our "Liberators"!

But that is another story.

Lech Laszkiewicz, F/O P-2316, retired from a career in aviation in December 1981, as a British Caledonian Captain. He lives in West Sussex.

❑❑❑❑

Pilot Officer Czeslaw Oberdak, P-2224, of 306 went missing on his 10th mission from Coolham, Ramrod 947, on 30 May 1944, when he was forced to make an emergency landing in Holland after his aircraft developed engine trouble. After hiding in various places he was arrested in December and believed to have been executed by Sicherheitsdienst (SD) on 8 March 1945, at Woeste Hoeve (Apeldoorn) when he was 23 years old.

This account of his likely fate was compiled by Dutch journalist Richard Schuurman who took up the story after P/O Oberdak's sister, Ludmila Kaczmarska, contacted the regional newspaper Zwolse Courant, the newspaper of the town of Zwolle, asking for help to find her brother.

P/O Czeslaw Oberdak, together with nine colleagues, took off from Coolham on Ramrod 947 at 9.20am on 30 May 1944. Their mission was to protect withdrawing US Air Force B-17 Fortressess from enemy aircraft between Nordhausen and Halberstadt over Germany.

The mission ended in disaster for him. The returning Fortresses and P-51 Mustangs met heavy flak at Schornebeck and Magdeburg and Oberdak's was probably hit.

He managed to fly a few hundred miles when over the eastern part of the Netherlands the engine developed trouble. He was unable to keep his Mustang III airborne and declared a "Mayday". At about 1pm he made a forced landing near the city of Zwolle and radioed his colleagues that he was Okay.

It was the last they heard from him. Oberdak was reported "Missing in action" and today officially he still is.

His family received their last letter from him after Christmas 1943. In Poland, his sister, Ludmila, had wondered for years what had happened to her younger brother.

After the war she had received some of his personal belongings but any requests for information to the RAF and the Red Cross remained unanswered. Her search was actually blocked when the Iron Curtain was raised.

But in 1991, a friend showed her a book on Polish pilots in which the events of her brother's last mission were listed.

The name of the town where Oberdak could have been (Zwolle) gave her hope and she decided to contact the regional newspaper, *Zwolse Courant,* asking for help to find her brother.

It happened to be me who read the letter first and I was immediately struck by its story.

With the help of the right people I was able to reconstruct in barely a month most of Czeslaw Oberdak's whereabouts in Holland since his emergency landing.

On her 74th birthday in September 1991 I visited Mrs Kaczmarska to tell her about her brother.

After he landed near the village of Dalmsholte, just east of Zwolle, Oberdak quickly took off his pilot suit and set fire to his Mustang aircraft before trying to hide in the woods.

Fortunately, he was spotted by the Resistance who took him to other pilots they were sheltering and the small group was moved to various hiding places around Zwolle, Harderwijk and, for some reason, even to Amsterdam, although the Dutch capital was full of Germans.

The local Resistance managed to find some bikes that had not been confiscated by the Nazis and on 6 December took a group of four pilots from Amsterdam to Otterloo, a small town in the middle of the country.

It was a risky journey that took a few days but they made it. As Otterloo was not that far from the river Rhine it was obvious that the pilots wanted to cross the river into the almost liberated southern part of the Netherlands and return through Belgium to their bases in Britain.

In Otterloo the pilots were hidden in a purposely made underground cave which had beds, food and even weapons. It was safe, until 24 December when they were discovered.

By pure coincidence a German truck had got a punctured tyre on the road nearby the cave. Probably because the pilots were making too much noise, they were discovered and arrested.

They tried to escape but Oberdak and American navigator Franklin D. Coslett, with whom Oberdak had been in hiding since his emergency landing in May, went the wrong way and were arrested again.

Coslett and Oberdak were immediately brought to a prison in Velp near Arnhem where, according to Coslett, they were tried and sentenced to death for espionage and terrorism. Late 1944/early 1945 they were transferred to the bigger Kruisberg prison in nearby Doetinchem.

It was in this prison that Czeslaw Oberdak was seen for the last time. What happened to him since has not been confirmed, but Coslett (who died in 1992) told me in 1991 that he had learned that "one night Czeslaw was taken away together with 40 prisoners and was executed".

I learned there had been two groups from the Kruisberg prison that had been executed. Of the first group, executed on 2 March 1945, all names are known and Oberdak's is not one of them. Around the second group some questionmarks still exist.

As a revenge for a failed attack on high ranking German police general Hans Rauter on 5 March at Woeste Hoeve near Apeldoorn, the Germans ordered 117 prisoners to be executed at the same spot.

In the early morning of 8 March, 117 men were taken from various prisons from all over the country to Woeste Hoeve and executed in small groups. Among them were 25 prisoners from the Kruisberg.

The 117 were buried in a mass grave in Ugchelen, but after the war Dutch police began their identification and reburial. As they completed their work, they realised they had a problem: of the 117 they were able to identify only 115. The two unknown remained at Ugchelen until 1982 when they were transferred to the war cemetery of Loenen. Each one is buried there as "Unknown Dutchman".

With Oberdak missing and believed to have been among the 117 and with two men officially unidentified from the same execution, I tried to find out if one of those two could be Oberdak.

Lt Harrie Jongen from the Dutch Royal Army war grave services, a specialist in doing identifications, had the records on the 1982 reburial of the two men, in his files. They showed the first man was too big to be Oberdak, but with an estimated height of 1.67 metres the second one came close to Oberdak's 1.63 metres (64.5 inches). According to Jongen, the skull was also typical for a man from eastern Europe.

Jongen took on the job to make an identification but first needed as much information on Oberdak as it was possible to get.

For this the medical documents on Oberdak that existed in the RAF archives were of absolute necessity but unfortunately it took more than two years for Jongen to get them.

At first, Lt Jongen tried to do an identification "on paper" by comparing data of Oberdak with that from the unknown man. As this was inconclusive, on 5 September

1995, he decided to dig up the remains for a more thorough investigation.

This might even include a DNA investigation that should settle the matter, hopefully in early 1996.

Taking everything into account, I would be surprised if the unknown man appeared not to be Oberdak as other options concerning the execution of people from the Kruisberg prison are very remote.

In order not to have him forgotten, Ludmila Kaczmarska, some years ago, had the name of her brother inscribed on their parents' grave in Krakow. She is also happy to know that Czeslaw is remembered in Coolham but her biggest wish is to know where he is buried. Hopefully, the answer is due.

❏❏❏❏

Flight Sergeant Franciszek Tomczak of 306 recalls:

On 1 April 1944, I flew with 306 Polish Squadron from RAF Heston to No. 15 Advanced Landing Ground at Coolham.

This was to become our operational station for the next three months.

We were to change our aircraft from Spitfire Mk IX's to Mustang Mk III's, which we were to ferry from RAF Aston Down.

On 4 April, after becoming airborne from Aston Down, I saw a friend, F/O Szajda, crash his aircraft on take off. He was trapped and could not be released. Despite all efforts, he perished with his aircraft.

It rained for days but we managed to get in some operational training and familiarised ourselves with our new aircraft.

On 21 April, whilst on local flying, I saw one of our new pilots, F/Sgt Osieleniec, coming in to land, when his right undercarriage leg struck a tree stump.

He must have opened up to go round and the leg broke his flap. The aircraft flipped over onto its back and struck the ground. He was killed on impact.

On 25 April I was asked to fly an Auster III to pick up someone from a south coast airfield.

On taxiing along on the grass field I felt the aircraft constantly pulling left. Thinking it might be the rough ground I tried to counteract this by applying the right brake. I lined up for take off and opened up full throttle, still holding with the right brake.

After rolling a few yards I had to release the right brake, when suddenly the aircraft turned 90 degrees left and about 50 yards ahead I was confronted by a mound of earth about six feet high.

I just had time to close the throttle and switch off the engine when the aircraft struck the mound and turned on its back.

I was strapped into the seat upside down when I saw petrol pouring down between my legs.

Realising that it might catch fire I undid the straps and crawled back inside the fuselage, trying to get out, but was unable to. I then looked back, and seeing no fire, made my way back to the cockpit where I managed to kick the door out and get out.

By then a number of people had arrived and my Wing Commander asked me what had happened. I told him the left wheel had siezed up on take off and he climbed up and tried it but it was stuck solid and would not budge.

I returned to my tent and within half an hour the pains in my limbs had become a stiffness which immobilised my entire body.

The medical orderlies picked me up and carried me to Sick Quarters where our squadron doctor, F/O Harasymowicz, treated me for three days.

I was given a few days' leave which took me into May when my second experience of hay fever started (the first being the previous year).

It was a bad bout and I was sent to hospital on 4 May.

As it was the end of my first operational tour I was posted to Blackpool Polish Air Force Depot on 25 May, so ended my stay at RAF Coolham.

Franciszek Tomczak, F/Sgt P-792015 made a career in the RAF and became a Flight Lieutenant. He changed his name to Tom Pavey and lives in Somerset.

❏❏❏❏

Corporal Tadeusz Jakubowski was a fitter with 306 Squadron.

I worked in one of the blister hangars occupied by maintenance unit 6306 Service Echelon.

If I remember rightly the hangars were close to Thakeham Road with a farmer's gate close by.

I think the sleeping site I have pictures of must have been site No. 1 or 2 as we had not far to go for our grub.

At some time we had to pack up and move to another site.

Our stay at Coolham was a short one and my memory of that time is small.

Tadeusz Jakubowski, Cpl P-781490
Cheshire.

❏❏❏❏

Ground crews obviously played a vital role, keeping the planes airworthy. They built up a rapport with the pilots who trusted them completely. One of 306's ground crew was Corporal Michael Tanner-Garbarz, who was an armourer. But before even arriving in Britain and eventually being posted to 306 Michael and fellow Polish airmen had survived a hair-raising journey through Europe after Hitler attacked Poland in 1939.

At the time he was a non-commissioned officer in the Polish Air Force training as an armourer near the Czechoslovakian border.

Trainees then moved to Luck near the Russian border. By the time they arrived Russia was also at war with Poland. More than 3,000 airmen were taken prisoner by the Russians, but when the coach carrying Michael's group of 22 broke down it saved them from a similar fate. A Polish border officer appeared and told them the safest place to go was Hungary.

After an 18-hour drive they arrived and the Polish Embassy in Budapest organised their passage to France through Yugoslavia and Italy. They arrived in France in January 1940 and were formed into Polish units attached to the French Air Force. But in June 1940 France collapsed and Michael was one of 63 who took lorries from the aerodrome at Rennes, in Brittany, where they were stationed – the French personnel just disappeared – and drove to Bordeaux. The American embassy provided them with a small boat and they sailed up through the Bay of Biscay, going in circles to avoid U-Boats, and then landed in Falmouth, in Cornwall, on 23 June 1940. Greeted by British Red Cross workers, after rest and training they were dispersed to different Polish units in various parts of the country.

These are his memories of the time he spent with 306 Squadron around the D-Day period.

The 306 (Polish) Fighter Squadron represented the town of Torun, hence the squadron in Polish was called Torunski.

But shortly before we were drafted to 306 we were stationed at Heston, near London, with 317 Squadron which represented the town of Wilno in the north east of Poland.

At Heston we started intensive training, preparing for D-Day. Everybody had to learn to drive but ground crew – mechanics, armourers and electricians – didn't have much time to be trained as qualified drivers but we had to learn how to drive a vehicle in an emergency, even if only in first or second gear.

But the other crew, who were only helpers to the mechanics, armourers and electricians, were trained fully to qualify as drivers of HGV vehicles.

At the time we were issued with Thorneycroft lorries. We were mobile, all our tools, personal belongings and tents were neatly stored in the lorries so that as soon as orders came to move out we would be ready.

We were longing for those orders, longing for the time when we could land in France or Belgium. Our orders came on 31 March 1944, to move out the next day.

We arrived in Coolham but without our pilots. They remained with their Spitfires and we were renamed 306 Squadron with Mustangs.

As soon as the tents were up, and with all the squadron assembled, we were introduced to S/Ldr Lapka and the rest of the flying personnel.

After a short speech by S/Ldr Lapka we started to get acquainted with the Mustang fighter/bombers. Each tradesman was alloted his own duties and mine were guns, bombs and ammunition.

There was no problem as we were already experienced with guns and bombs in Poland and France and with the RAF Hurricanes and Spitfires. We even had knowledge of the 20mm Hispano-Suiza cannon, so in a matter of hours we were familiar with the equipment.

It took a bit longer with the bombs as we had to load them by hand, two 500 pounders to each aircraft. It took four men to lift the bomb up into the bomb-rack, pushing up until we heard the click, then securing the bomb with four adjustable screws.

My job was then to screw in the fuses with a split pin attached to the bomb-rack. When the pilot was flying and aiming at the target all he had to do was press the button.

Two bombs, one from the port wing and one from the starboard, were released at the same time, the split pin was pulled out of the fuse and the bomb was ready to explode when it hit the target.

Our first three weeks in Coolham we didn't get much practice as it was raining non stop. But when the weather improved our pilots started the "softening up" operations, attacking German positions in France, usually four times a day.

As soon as it became light, just before sunrise, we were working on our aircraft making sure they were ready for action and signing Form 700.

This meant the aircraft had been guaranteed by the mechanic, armourer, electrician and instrument maker and the pilot had to sign to show he had examined the aircraft and was satisfied.

I very seldom saw pilots doing this, they had such confidence in the ground crew, they trusted us implicitly.

Of course, they never experienced the failure of an engine or guns or bombs not dropping off when they pressed the button while on missions over France.

The sad thing was waiting for our pilots to return, counting them in, with some occasionally failing to return having been killed in action.

Some also crashed when taking off or landing. One pilot from 129 Squadron hit the trees shortly after taking off. We cycled over to see his body and wreckage scattered in the woods. He was a young British chap (F/O Stanley Payne).

All the pilots from 129 Squadron were British or from the Commonwealth nations but the ground crews were Polish.

Sgt Mackowski was in charge of armaments for 129 Squadron. Incidentally, we had trained together at the beginning of 1940 on Hurricanes and Spitfires.

Often, ground crews worked from early morning before sunrise until dark. But we had a rota, giving us two or three hours off about twice a week, so we took the opportunity to explore the countryside.

Some of the boys liked to go dancing, I did like a pint and at the same time the chance to meet some British people. My favourite spot to have a pint was the Queen's Head at West Chiltington and sometimes the Rising Sun at Nutbourne.

Easter came and we had an invasion of schoolboys who were very interested in our Mustangs. I soon became friendly with Barry Lenhart, who was just 11 at the time.

He was at the airfield very early in the morning, before going to school. I shared my breakfast with him as we were getting more food than we could manage to eat.

He was so happy I let him watch me when I was busy working with the guns and bombs.

After a week or two I had a letter from Barry's mum and dad asking if I could come to tea one Sunday at their house in West Chiltington.

Our friendship developed from there and I used to spend my leave with the family right up until the end of my time in the RAF.

Our days in Coolham became very busy at the end of April, and May was even busier.

In the middle of May the American army held an exercise involving a mock attack on our airfield. Unfortunately, someone had made a mistake!

About 30 Dakotas, each one pulling a glider, with approximately 40 troops on board, approached Coolham airfield but the order to release the gliders came a few seconds too late and all the gliders landed in Shipley, in gardens and orchards, many of them breaking up against the trees.

We had orders to jump on our bikes and go there quickly to help retrieve the injured soldiers.

It was an ugly sight, the injured crying with pain.

By the time we had managed to do what we could to help, our own aircraft were coming back from operations over France and we had to rush back to refuel and reload the guns and bombs. Our shirts were wet with sweat.

When D-Day arrived we were in a state of euphoria, at last the time had come.

Pilots coming back from action over France didn't want to waste time, they only got out of their Mustangs just to have a cup of tea or a small sandwich while the aircraft were re-armed and refuelled and the engines checked, then they were off again.

Operation Overlord was in full swing when we experienced the first surprise V1 rockets late in the evening. The alarm hooter sounded and then we saw the first flying bomb.

We were expecting something like this, the secret weapon, long before, but it was a strange feeling seeing one flying just over our airfield.

From that day on we saw many and our pilots got to know them and how to shoot them down before they got to London but unfortunately many of them did get to London.

The majority of us didn't take much notice of them knowing that when the motor was working the V1 would continue flying on but as soon as the motor cut out we were quick to take cover.

But we did have one casualty, an armourer, who suffered from shell shock and had to be taken to a mental hospital.

That was at the end of June, beginning of July, when our pilots and the anti aircraft defence on the coast became so good at shooting the V1's down that hardly any got through. However, unfortunately for us they were coming down anywhere and sometimes close to the airfield.

That was the time when the chap's nerves couldn't take it. We felt sorry for him, it was a pitiful sight, but he was taken away so the rest of the section would not be demoralized.

I just shudder to think what they would have done with him in the First World War (cowardice – firing squad).

Once the Allied Forces were firmly established in France our Mustangs were fitted with spare petrol tanks instead of bombs and were escorting bombers deep into Germany to bomb important targets.

Those missions were very successful and the Mustangs were well suited to this type of operational flying which took about six hours.

The ground crew were as usual anxiously waiting for our pilots to return, happy when we counted them all safely back in.

It was a time when we all worked hard, there was danger every minute of the day and night, for the pilots during their missions and for us during attacks by V1's.

Soon, American Dakotas started landing at Coolham, bringing German prisoners of war who were put in a small compound in Shipley for a day or two and then they were moved to larger camps further north.

It was shortly after this that the squadron left Coolham for Holmsley South in Hampshire then Brenzett in Kent where our pilots were involved in shooting down V1's as they flew in three lines over our airfield every 15 minutes.

Eventually, after a six month stay in hospital following an operation after I had the bad luck to get osteomyelitis in my left leg when the cold penetrated right through to the bone, I was transferred from 306 Squadron to 411 RSU on the continent.

The war came to an end and we, the Polish Air Force, had our most bitter disappointment. Poland was under communist control and, of course, we had to think and plan where to settle down.

After postings in Germany with BAOR and RAF Kinloss in Scotland I was demobbed in November, 1947.

We did not choose to fight, to satisfy man's animal instinct to kill, but once attacked we had to defend our land, our families, our freedom, and being ready to give our lives, we have done so for our freedom and yours.

Michael Tanner-Garbarz, Cpl P-782314, married his Belgian wife, Lucienne, in October 1947. He returned with her to West Sussex where they have lived ever since.

Corporal Michael Tanner-Garbarz
Photo: M. Tanner-Garbarz

306 (Polish) Squadron. *Photo: Mrs J Sologub*

F/Lt Wladyslaw Potocki, DFC (left) and F/Lt Henryk Pietrzak, DFC. *Photo: Mrs L. Pietrzak*

F/Lt Lech Laszkiewicz. *Photo: L. Laszkiewicz*

306 Squadron pilots at rest, beside the volley ball net, Coolham ALG. From left: P/O Kawnik, F/Sgt Krzehlik, F/O Porazinski, F/Lt Pietrzak, F/O Tronczynski, P/O Smigielski. *Photo: Z. Kawnik*

Primitive field conditions endured by 306 Squadron mechanics. *Photo: S. Gogolowski*

A few lighthearted instructions for 133 Wing Chaplain, S/Ldr Conrad Stolarek, seated in jeep, from Polish mechanics.

Personnel from 306 Squadron relaxing at Coolham. Clockwise from left: F/Lt Potocki, P/O Smigielski, F/O Tomszyk, F/Lt Pietrzak, F/Lt Siekierski, S/Ldr Lapka. *Photo: W. Krupa*

Polish P-51c Mustangs of 133 Wing commence another mission from Coolham. *Photo: J. Polak*

8

411 (Polish) Repair and Salvage Unit

Flight Lieutenant Karol Sienkiewicz was equipment officer with 411 RSU.

While the war was in full swing I remember being sanctioned by W/Cdr Carle, our group senior equipment officer, to acquire various extraordinary items of equipment suitable for married quarters and not for the life in tents. It included washhand basin, bath, shower, taps and some other equipment for fitment in a specially constructed caravan.

When the caravan was complete it was presented to Viscount Montgomery of Alamein and I believe it was brought to England after the war and is still at the late Viscount Montgomery's estate.

I do remember also taking part in a football match but scored no goals.

My social participation was very minimal. I stayed at the Red House guest house (Butterstocks Lane, Shipley) with my wife and we spent some evenings with R. Dabrowski, C. Komocki, Matey (Mrs Maitland), S. Skalski, Mrs Taytham and some others.

I stayed at the Red House (now Blenheim Cottage) practically all the time while in Coolham.

The in-laws even brought our only son, Michael, then two years old, who was craving for his mother.

S/Ldr Colin Maitland, DFC, died in 1945. His mother joined ENSA as a driver and visited her son's grave in Germany some time in 1945.

I met her in Bruges just before the end of the war.

After the war she settled with her mother in Littlehampton. We visited them back in the sixties and I have a picture of Colin in front of the main entrance to the cottage in Shipley.

Karol Sienkiewicz, F/Lt P-2337
Scotland.

Corporal Stanislaw Gogolowski, a mechanic, said 411 RSU was moved to RAF Coolham on 1 April 1944.

Our task was to repair and/or salvage all fighter planes damaged in action.

Our squadrons at Coolham had the task of softening the enemy's resistance before the invasion of Europe.

The RAF successes across the Channel had been quoted on the radio and in the press, telling of destroying supply trains, ammunition and petrol dumps and other important targets.

Mustangs were very successful flying in pairs, they were very good in stopping doodlebugs over the Channel before they could reach the English coast.

The tactics used by some pilots were very risky. They were lifting a doodlebug wingtip with their wing, then the bug lost its balance and down it went.

On the airfield at Coolham there was a lake with plenty of water birds which were troubled by pikes but our cooks very obligingly caught the pikes bringing relief to the ducklings.

We also had some break from boredom having to shake black snakes out of our best Sunday blue before going to the village dance.

I also recollect seeing a windmill on the east side of the airfield.

One day we found a few American troop gliders in the woods nearby which could not make it to our landing strip. There were no lives lost.

Just before the invasion I was posted back to Bomber Command in Lincolnshire and lost contact with 411 RSU.

Stanislaw Gogolowski, Cpl P-780802
Lincoln.

Meals outside on fine days sampled here by S/Ldr Dabrowski (left), C/O of 411 RSU, and other personnel. *Photo: S Gogolowski*

Cpl Gogolowski (second left) with personnel of 411 RSU outside camouflaged workshop at Coolham. *Photo: S Gogolowski*

Personnel of 411 RSU at work. *Photo: S Gogolowski*

Groundcrews from 133 Wing prepare to move from Coolham on 22 June 1944. *Photo: W Krupa*

9

5090c (Polish) Mobile Signal Unit

Top secret messages are recalled by Leading Aircraftman Stanislaw Czajkowski, a radio operator.

I was 19 when stationed at Coolham with 5090c MSU which was a small unit responsible for radio communications.

We were attached to HQ and the unit consisted of radio transmitters, receiver van, generator van, cypher van, two lorries and two motorcycles for dispatch riders.

We slept in tents, six in each, on the other side of the road.

I remember the weather was wet and cold and even the invasion had to be postponed, also the grass was very tall on that field.

We worked in three shifts of eight hours each, keeping the radio section working 24 hours.

One day transport was provided for personnel wishing to go dancing in Horsham. I went and enjoyed every minute and, of course, we decided to escort girls home after the dance hoping to get a goodnight kiss.

Naturally we missed the 11 o'clock transport back and we had to walk back arriving just in time for early morning shift (vive l'amour!!).

There was an incident when the pilot of a crashed aircraft was killed and a farmer crying for his damaged tree saplings planted by his father. I was very hurt at that.

I also recall four chaps from the RAF Regiment being drunk and accosting me and my friend walking to the cookhouse for afternoon tea. Some punches were exchanged and my friend suddenly disappeared but only to summon help.

Soon I could see F/Lt Swan, a portly, short man, running in front of the men from the unit and soon all four were escorted to the guardroom.

Next day the Sgt Major, with a big stick under his arm, arrived to apologise for the behaviour of his men, looking me up and down, not having a scratch on my face, while four RAF Regiment men had four black eyes. But it was not my doing. Anyway, all was forgiven in a good spirit.

This is just a brief description of happenings. We were busy all the time.

Squadrons were flying non stop and also soon Hitler started sending flying bombs on London.

I remember a high priority coded radio message which I received informing of the invasion. After that we were in constant readiness to go over which we did soon after.

The officer commanding our unit was F/Lt Jedryszek. Also attached to our unit was a cypher section consisting of F/Lt Swan and two English sergeants, one of whom was named Sgt Polland. They were responsible for coding and decoding messages.

It is nice to know that somebody wants to remember all our stories.

LAC Stanislaw Czajkowski, P-705259
Southampton.

5090c (Polish) Mobile Signal Unit

Back row, from left: LAC unknown (driver), LAC unknown, W/O Rudnicki, LAC Krowicki (radio operator), F/O Jedryszek, LAC unknown (depatach rider) LAC unknown, LAC unknown (electrician), LAC Sawicki (despatch rider)
Seated: LAC unknown, LAC Czajkowski (radio operator) LAC unknown (motor mechanic) *Photo: S. Czajkowski*

Camouflaged Thornycroft lorries used by 5090c MSU and by 133 Wing as a mobile medical centre.
Photo: M. Jarosz

10

129 (Mysore) RAF Squadron

This multi national squadron, with pilots from Britain, Allied and Commonwealth countries, took its name from the Maharajah of Mysore in India who supported it financially. He put up the money for the first complement of Spitfires and every two or three months had his Trade Commissioner in London supply funds for a "really good party".

I suppose it all began in February 1944 when 129 Squadron was "on rest" at Peterhead, recalls Flying Officer George Frederick Pyle.

Rumours began to circulate that we would soon be going south to convert to the North American P-51 Mustang. While there would be a certain sadness at leaving the Spitfire, the prospect of being able to do long range escort duties and deep penetration into Germany was stimulating.

Rumour became fact when, on 15 March the squadron left Peterhead for Heston in order to prepare for the necessary conversion. But it soon became clear that long range escort duties were not really what we were going to do.

Instead, we were to be transferred to 2nd Tactical Air Force in preparation for the coming invasion of Europe and be employed in attacking enemy ground targets with bomb and bullet. It was different, exciting, challenging but also – as we were to find out – dangerous!

Before even getting into a P-51 Mustang it seemed that we had to learn the rudimentary lessons of dive-bombing and for this the squadron was sent to RAF Llanbedr in North Wales.

On the day of departure my pal, Eddy Edwards and I, received confirmation of our commissions and instead of flying to Llanbedr we were given the day in which to stop in London and get fitted with uniforms and then to follow on the night train.

The actual bombing course was brief and basic. I recorded two dive-bombing practice flights which grossed 65 minutes and left me in ignorance of how close my bombs had come to the target.

There were also three ground strafing sorties, two in flight strength and one where the targets were derelict trucks. So much for preparation for the serious business ahead.

Having gone to Llanbedr by train, Eddy and I were detailed to return the same way, but instead of going back to Heston we were directed to the temporary airfield at Coolham, near Horsham, which was to be the Squadron's "jumping off ground" for the invasion.

Since bad weather had grounded the air party we arrived first with the ground party, and a dismal arrival it proved to be.

We knew little of our new role except that the squadron would be the only RAF squadron in 133 Wing which would otherwise comprise two Polish squadrons.

The airfield would be of a temporary nature with runways simply cut through existing fields and covered with metal tracking to prevent rutting.

Since the Wing was expected to move lock, stock and barrel as soon as further temporary airfields became available on the Continent after the invasion, it had to be mobile and to this end everything was going to be under canvas.

The ground party arrived at the end of a long day during which it seemed to have rained endlessly and Ed and I found our new accommodation to be one of a line of tents sagging wetly at the edge of a recently cultivated field.

After a summer in permanent quarters at Hornchurch, even the icy Nissen huts of Peterhead had not prepared us for what now faced us.

Our tent contained basic officer's camping gear which included two camp beds upon which were stacked pillows and two blankets.

Everything was wet and even the blankets felt damp and chill.

The evening was gloomy and having made up our beds and stacked our gear as dry as possible there was little to be done except the obvious.

Where was the nearest pub? At least it would be light and dry. With vague directions from someone now long forgotten we set off with young Monty on his lead.

Monty was a fine little Alsatian pup, son of Jan Bush the squadron mascot and the property of P/O Mark Twomey who was still at Llanbedr. We were to find out that Monty was already beginning to suffer from distemper.

We found the pub, though I remember little of it, and eventually returned to the tent where I tied Monty to the tent pole. The floor was as the farmer had left it, muddy with long wet weeds and the odd cabbage stalk.

Feeling sadly disenchanted I settled into my bed, the damp blankets giving little warmth. Sleep did not come easily, Monty whined unhappily and finally, chilled to the marrow, I leapt out of bed with cramp in both feet.

Taking pity on the dog I unleased him, whereupon he promptly jumped up onto my bed and went to sleep on my feet. The resultant warmth was bliss and there were no more cramps. Perhaps that arrival at Coolham was the thing I recall above all.

Things improved out of all proportion when the sun finally shone and we began the job of settling in and making ourselves as comfortable as possible.

The air party had been delayed two days by bad weather and by the time they arrived things were drying out.

We had discovered that our "flights" were on the far side of the airfield behind a clump of trees close to a small lake. The flight approach to one of the runways came in over the lake and it had been necessary to fell some of the large trees which had bordered the airfield side.

Evidence of this was in the stumps which remained. Apart from the marquee which acted as the crew room, there was also the adjutant's tent, another for the C/O and another where the parachutes were stored.

To get from the "domestic" site to the "flights" the squadron had its own transport.

Messing was in a marquee close to the accommodation tents and it was there that we first met our Polish colleagues from 306 and 315 Squadrons.

It was not easy to make friends or conversation with them. The language barrier itself made communication difficult and the Poles were already an established "family" as it were.

They did not need the Commonwealth mix of 129, though they were courteous and shared the unique but unspoken brotherhood of the flier.

Perhaps for this reason some of us did not spend much time in the Mess where the usual drinking went on in the evenings but frequently, as bed time approached, we could hear our Polish comrades singing.

They had some good voices and a degree of harmony and the repertoire ranged from the noisy and patriotic (and doubtless bawdy at times) to the haunting tunes of their homeland.

It always finished with their National Anthem sung with fervour and pride. They were so far from the homes which few of them would ever know again.

We were not so impressed with their cooking, however. To our minds, good food was ruined in the kitchen even allowing for the field conditions, and we often left the Mess hungry.

Breakfast seemed to comprise cold, under-cooked tinned bacon and congealed fried eggs sticking to cold plates.

Even after almost five years of war, soused herring accompanied by rice wrapped in a cabbage leaf seemed a pretty poor substitute for a main evening meal!

We eventually got round to claiming fresh eggs, which, when fried in a billy can over a small fire outside the tent and served with fried bread and a packet of crisps, made a welcome evening snack.

Ablutions were fairly primitive with the canvas wash-basins of great value, but the miniscule canvas baths (officers for the use of) were not popular and but little used.

As time went by some chaps found their way to local homes for showers or baths; the remainder were able occasionally to visit a local army Mess. What luxury!

Dress was what could only be described as "scruff". We lived in our flying clothes, not the modern flying overall but simply battledress of serge slacks and button-up blouse. Silk or nylon scarves were preferred to ties since the heavy white sea jerseys were generally worn under the blouse, and we went everywhere in our flying boots.

The old suede, fleece lined "wellington" style flying boots which, while comfortable and warm for flying, were clumsy, bad for walking and had a habit of being jerked off in the event of a parachute descent, had long since been replaced by the "escape" boot.

This was simply a stout, black, fleece lined, lace up shoe with a lightly stitched, zip up legging with a strap around the top. In the event of a parachute jump the boots remained on the feet and upon landing, the leggings could be quickly removed and the pilot left with serviceable and inconspicuous walking shoes.

Such was the theory, but certainly they were worn around the airfield when the old ones might not have been.

We had been issued with side-arms; the black, break type Webley .38 but these were little worn until the invasion started.

Meanwhile, everyone was anxious to try the Mustang. There was instruction in the petrol, oil and cooling systems; we studied the flying and handling instructions and spent as much time as possible in the cockpit of any available aircraft.

Throughout those first few days flying conditions were poor, if not prohibitive. One by one the old Spitfire Vs which we had brought down from Peterhead began to disappear, collected by a mixture of RAF ferry crews and Air Transport Auxiliaries.

At last the weather cleared and everybody got their first flight on type. Unlike my first solo or my first Spitfire flight, I do not recall much of the occasion except for the feeling of vulnerability.

The cramped, high sided cockpit of the Spit had given a feeling (utterly false) of protection. There in the Mustang, with its large, sliding Malcolm hood, I felt exposed from the waist up.

The visibility, in spite of the long nose, was excellent;

the pistol grip control column seemed strange after the spade grip of the Spit and the banks of switches and buttons were strange, but the Packard Merlin engine was quite familiar.

Above all, the extensive flaps and wide track undercarriage were a delight after the narrow track of the Spit.

My log book records "Experience on type – four landings" and flying the Mustang was indeed an experience.

Every pilot registered pleasure and enthusiasm and there was at once speculation as to when we would become operational.

Many thoughts were directed at the possibility of a low level "Ranger" against enemy airfields or selected ground targets but first there was need for more work to be done.

The urgent business of flying training was pushed as hard as weather, ground conditions and aircraft serviceability permitted and we found ourselves involved in formation flying, low level cross country flights, army co-operation exercises and practice dog fights.

On one local flight, Eddy and I were "bounced" by a couple of optimistic Typhoon pilots and the most hectic of dogfights ensued. In fact, we saw them coming and were astonished at the way we were able, eventually, not only to avoid their persistent attempts to take us but to reverse the roles and sit on their tails all the way back to their base at Tangmere.

We returned to Coolham with a new respect for, and confidence in, the Mustang.

The new aircraft arriving each day were of the RAF specification with the large sliding Malcolm, or "bubble" hood, which, as already mentioned, afforded excellent visibility.

Unhappily, the aircraft bearing my squadron letter "B" was of the older type which still had the fold-over, flat panel cockpit cover.

Visibility was dreadful, particularly to the rear and it had a definite claustrophobic effect.

There could be difficulties if there was a need to abandon the aircraft in a hurry. "B" Baker also seemed prone to unserviceability.

I looked forward to receiving the new type promised by my flight commander "as soon as possible". In the event, I never did see the new one and did most of my flying in whichever other Mustang of the later type was available at the time.

And still the squadron remained non-operational, until, that is, 26 April when a flight of four aircraft took off on a Ranger into northern France.

It was an "A" Flight op. led by Joe Leigh, our flight commander, and supported by Des Byrne, Guns Conroy and Rocky Nelson.

The target was the Luftwaffe fighter base at Beauvais and the hope was that they would catch a few fat FW190's on the ground. The general feeling among the envious pilots left behind was "lucky sods".

Some time later when only three aircraft touched down there began to be misgivings, followed later by dismay and real sorrow when the facts came to life.

Flying low to avoid detection, the flight had arrived on target only to find an empty airfield and an alert airfield defence.

The flak had been intense, almost as though they had been expected, and three of the four aircraft were hit. Apart from firing at secondary targets on the airfield perimeter, the attack yielded little, but it cost dearly.

Des and Guns were nicked by flak but continued. Rocky, however, must have been hit severely and as he dropped back was heard to say something about his Chinaman's luck and that he was getting hot.

He disappeared from view but a cloud of black smoke mushroomed out of a wood. His loss left another cloud over the squadron. It had not been an auspicious start to operations on the type.

My own first Mustang op. was three days later when, flying No. 2 to the C/O the squadron carried out a Ramrod in poor weather conditions. Flying just below cloud at about 3,000 feet over NE France, we were subject to intense flak and I knew my aircraft had been hit, though I had felt little.

There were no signs of damage coming from my instruments and power was normal but I watched the radiator temperature gauge carefully in case I was losing Glycol coolant.

In fact the only damage turned out to be a small hole in the radiator scoop caused by a raggy splinter still lodged in the fairing. Later in the day I was one of six Mustangs which flew to Fairwood Common where we refuelled before flying west to rendezvous with a Dakota carrying the Australian Prime Minister to London.

The escort to Croydon involved a flight of three hours which gave us some idea of the duration of this long range fighter. It also brought my flying time for the day to five hours twenty minutes.

It was on a later Wing operation led by Group Captain Nowierski that we lost two Mustangs before the Wing was even airborne.

Earlier in the day one of the Poles had landed with his wheels up and in doing so managed to tear up a good length of the metal tracking.

By the time the Wing was ready to take off in the afternoon the runway was still unserviceable and take off was at right angles to a powerful cross wind.

Each aircraft carried two 500lb bombs. The two Polish squadrons took off first in pairs, raising a bit of dust as they did so and then, with the Group Captain leading 129, we began our take off.

First went the Groupy with his No. 2, followed by me with Neil Roggenkamp as my No. 2 and then the next four pairs.

When airborne I had some difficulty finding the Groupy's pair up front and with the rest of 129 formation upon me I finally spotted the Wing leader as a single aircraft.

I was also puzzled at the absence of Roggy who should have been tucking in alongside me.

It was a good operation with good bombing amongst the railway stock in a small marshalling yard but the mystery of the missing Mustangs remained until we landed.

It transpired that on take off the Groupy's No 2 had swung violently to port in the strong crosswind, his undercarriage had collapsed and he had gone sliding into a field latrine, scattering those nearby.

As the ground crews then rushed to the aircraft they were promptly scattered again as Roggy did exactly the same at almost the same spot.

Fortunately none of the bombs went off and neither aircraft caught fire. Including the aircraft that had landed in the morning with its wheels up, the Wing suddenly had three damaged and quite unserviceable Mustangs in one day without a shot from the enemy.

It has been said that war is 90 per cent boredom and 10 per cent action and certainly life on a squadron bore this out to a great extent.

Everyone wanted to fly, either operationally or just to be in the air, and every excuse was used to do so, whether for practice or just an air test.

Operations had not reached the hectic stage. The result was that a lot of time was spent hanging about on the ground waiting for the call into the air.

The pilots could only spend so much time checking or cleaning their aircraft; there were card schools, discussions and the occasional bit of training, but generally time could drag.

Since boyhood I had had a passion for flying and all things appertaining to it and I could spend hours happily watching aircraft taking off and landing.

One sunny day I was watching one of the old Spitfires being collected and as it trundled down the runway on its take off run I wondered why the pilot was taking such a long time to get airborne.

It then became apparent that he was still in coarse pitch, apparent, that is, to everybody but the pilot.

He finally clawed his way into the air and started a turn to port, but when at about 400 feet the inevitable happened and we all watched as the Spit stalled and fell to earth behind some trees.

A number of us jumped into the truck and followed the crash wagon down the road where it stopped by a wood. The Spit had come down in a huge briar patch in the wood.

Its back was broken although it was right way up and we were in time to see the pilot being stretchered away in a somewhat bloody state. We never did know the extent of his injuries.

On another occasion as one of the Polish Mustangs approached the runway from over the little lake I heard the urgent sound of his throttle being slammed open and was in time to see it rearing into the air from a cloud of dust in order to abort the landing and "go round again".

As it staggered away gaining speed and altitude I realised that his starboard undercarriage and what appeared to be his starboard flap had disappeared.

With the loss of lift on the starboard wing and the extended flap still down on the other the pilot was quite unable to hold it and as if in slow motion we all watched as the aircraft half rolled and crashed inverted at the other end of the runway.

The poor chap had landed short and although he had cleared the lake, sadly he had struck the stump of one of the trees which had been felled to clear the final approach.

The undercarriage leg and wheel had been ripped off and had carried the flap with it. The pilot's only hope of survival would have been to select the remaining leg up and to continue the landing on his belly, but such was the speed of events he would not realise the extent of the damage in time to make such an instant decision.

As pilots we were never really used to pals going missing but it was accepted. To see the like of this right on the airfield had a particularly sobering effect upon almost everyone. (Incident recollected refers to the death of F/Sgt Osieleniec of 306 Squadron.)

A similar tragedy which occurred in my absence and which appears to have been one of those little oversights, was the death of F/O Payne – Skip.

From what I was told afterwards it appears that he was taking off with the squadron on a bombing sortie and just as he was climbing away from the runway his engine died, he lost height, clipped the trees and crashed in a ball of flame. It could have been engine trouble but the accepted theory was that he had failed to tighten the friction nut on the throttle quadrant (usually slackened off when taxiing) and as he took his hand off the throttle in order to raise the undercarriage, the throttle had slipped back at the moment when he needed maximum power.

Another non-operational accident occurred to one of the Poles who, it seemed, had failed to pull out of a high speed dive. It was reported that all they ever found of the pilot was a piece of battle dress in a tree, but such tales may be exaggerated to emphasise the horror. He certainly died.

By the beginning of May the squadron had seen a few changes in its make up of pilots.

Several of the "old" faces had been posted out either on rest or to other duties and been replaced by new ones.

In addition, policy had decided that a normal complement of 24 pilots should be cut to 18 and W/O Cliff (an ex glider pilot who had been with 129 since the previous summer), P/O Edwards and F/Lt Green were posted to 84 GSU, a holding unit at Aston Down.

The idea was that casualties would be replaced by these reserves who would fly in with a replacement aircraft and immediately fit smoothly and effectively into the squadron which they knew.

Those selected were old hands. Eddy, Mark Twomey and I had joined 129 together a year previously and I suppose I was fortunate not to be one of them.

Newer faces included those of P/O Humphries, F/Lt Kleimeyer and F/O Thomson.

By the time I returned to the squadron in September there had been more changes still; and more losses.

The backbone of the squadron was the C/O, S/Ldr Wag Haw, who had taken over in the previous autumn, supported by his two flight commanders, F/Lt A.C. Leigh (Joe) and F/Lt A.J. Hancock, and it carried a depth of pilots with F/O Les Lunn and F/O G.R. Dickson probably having been with 129 the longest.

"Dickie" had been lost over Normandy the previous summer and had made one of the swiftest evasions (avoiding capture) over the Pyrenees to return to the squadron.

The "Ace" was probably F/Lt Ruchwaldy (Rush) who had had a very successful summer of 1943 with several confirmed victories. He had been born in Malaya and maintained that his home was the first to be destroyed by Japanese action.

By the time he arrived at Coolham he was the next in line to be a flight commander and indeed he was to command "B" Flight after Alan Hancock's death.

The squadron was like many other wartime squadrons in that it was made up of pilots from Britain with a strong Dominion content.

The Australians were represented by F/O "Guns" Conroy, W/O "Johnnie" Byrne, W/O "Tommy" Thomas, W/O "Joe" Foster and W/O Neil "Roggy" Roggenkamp.

They had all been with 129 since the previous summer at Hornchurch. Byrne was to return to Australia before the Second Front opened and Roggenkamp was to do likewise in August after his return to the UK following a successful escape back.

Conroy remained with 129 until sent on rest at the end of 1944 when he was posted as an instructor to the Mustang OTU at Rednall (as did several other 129 types).

F/O "Dicky" Dickson was a New Zealander and was to meet his wife while stationed at Coolham.

I was always of the opinion that F/Lt Kleimeyer was also a Kiwi but he could have been an Australian who had been in the NZ Air Force. Likewise, I felt that F/Lt "Stalky" Strachan was Canadian though some say he too was an Aussie.

F/O "Tommy" Thomson certainly was a Canadian, as "Rocky" Nelson had been. F/Lt "Seaweed" Baker was American and died in mysterious circumstances in London while on leave. All in all it was a friendly mob with the lads from Dominions bringing their own inimitable high spirited humour and behaviour to the squadron.

Backing up the pilots were those who served without the "glamour" of flight or conflict but without whom a squadron could not exist. Strangely enough their names have mostly slipped from memory.

Perhaps the most important was "Daks", the Engineering Officer, who, working with Polish ground crews, was responsible for the serviceability of our aircraft.

His were the problems of converting onto the Mustang, accepting new aircraft as they arrived, training his crews, seeing that aircraft were armed, fuelled and inspected daily and of ensuring that reported faults were investigated and corrected.

All of this was generally done out in the open in all weathers. Another of his problems was the loss of aircraft to normal duties as major inspections became due, or damaged aircraft were repaired or sent to a Maintenance Unit.

The new Adjutant was F/O Madge – "Madge the Adj" – a tall, lean, dark haired man who was understood to be a fast bowler who had played for Essex.

He kept the wheels of administration oiled and turning with responsibilities ranging from postings and travel warrants to squadron diary and requisitions.

Nobody really knew what he did except that if something had to be arranged, obtained, disposed of, reported, catalogued or was lost, then he did it.

The Squadron Medical Officer was "Doc" Mason, a quiet, serious young man whose opposite numbers on the other squadrons were Poles.

There was also the Padre, "Mac" the Equipment Officer and "Gibby" the Gas Officer.

The Intelligence Officer had been F/O "Chappy" Chapman, a peace time movie actor whom I had seen playing opposite Raymond Massey in H.G. Wells' *The shape of things to come.*
I cannot recall when he left the squadron and am unable to say just who the Intelligence Officer was at Coolham.

Jan Bush was the squadron mascot and had initially been the pet of F/O Bush, one of the early pilots. Unfortunately, he had been killed in a flying accident – the story that I heard was that his seat type dinghy had inflated in flight, pushing the control column forward and preventing Bush from pulling out of the ensuing dive.

Responsibility for Jan had then been passed on to F/Sgt – later F/O – Tony Bradshaw, but in December 1943 Bradshaw and his No. 2, F/O Nigel Cain, collided in mid air over northern France.

Cain died and after a brief attempt at evading the enemy, Bradshaw was captured and spent the rest of the war in a prison camp.

The next "minder" was Skip Payne and we know the nature of his unfortunate demise. It seemed that Jan, a heavy, thick coated Alsatian, was a jinx on anyone looking after her and I never knew which courageous soul accepted the responsibility after Skip's death.

One characteristic of Jan was that she resented strangers on the site unless they were wearing flying boots.

Some time after the war I saw a photograph in a national newspaper of a number of Royal Auxiliary Air Force pilots running in a mock "scramble" to their Spitfires and in the group could be seen Tony Bradshaw with Jan Bush leaping at his side. Perhaps the dog ended her days with Tony.

Of occurences outside the airfield there were not many during my stay. Weather permitting, Les Lunn and I (Eddy had gone to Aston Down) would go walking in the surrounding countryside, especially in the warm summer evenings. One such evening was particularly memorable.

It had been a day of intense activity by both the US Army and Air Force in a combined practice airborne drop by gliders. We had seen the towing aircraft flying over and releasing their gliders in the surrounding countryside.

In the evening, Les and I went for one of our walks and were quite shaken to see the condition of some of the gliders, mostly Hadrians, though there were some Horsas.

As pilots, we looked upon a good landing as one in which neither pilot nor aircraft suffered damage or shock, NOT one from which one simply managed to walk away.

The gliders on the other hand seemed to have been thrown down anywhere and they lay around smashed and broken, some in the open, but less fortunate ones among the trees.

One Hadrian had come to an abrupt halt and it appeared that the vehicle it had been carrying had broken loose on impact and smashed its way out of the glider via the cockpit and nose.

The thought of the casualities which must have been suffered was chilling; and this had only been a practice operation. In the real thing, how many men were going to be able to take part in the battle?

It may have been later in the same evening that Les and I were walking in the fields when a thick mist came down. We became hopelessly lost and actually began to feel a certain panic as time went by and dusk began to make things worse.

We eventually found our way back to the field but, like seeing the gliders, it had been a sobering experience and not much of a recommendation for ground navigation in the event of landing in enemy occupied territory.

A more pleasant experience followed when we were approached by the Padre who explained that he, Mac (Equipment) and Gibby (Gas), had spent the previous Saturday evening as the guests of a Mr and Mrs Hales at their duck farm just up the road (now St Andrews Farm boarding kennels and cattery, Brooks Green).

Mr Hales had asked the Padre if he would invite a couple of the pilots to come along too on the following Saturday. In his infinite wisdom the Padre had obviously recognised in Les and me a couple of fine types who would not disgrace themselves or the Air Force and really, not knowing what we were letting ourselves in for, we accepted. And glad we were that we did!

When we arrived at the farm we were introduced to our host and led to a large wooden summerhouse in the garden and then introduced to five very nice young ladies.

The floor of the summerhouse was cleared and polished for dancing and at the end was a little bar and record player or radiogram. We were soon at our ease with our new friends and proceeded to dance the night away.

Half way through the evening Mrs Hales arrived with coffee, sandwiches and a large cream cake. What luxury!

From then on the same little group spent each Saturday evening with our new friends and I very much doubt if our hosts could possibly appreciate what an enormous fillip it gave to us to dress in our best uniform, shake off not only the scruff of canvas living but, for a while, the tensions of operational flying, and spend a happy and relaxed time with friends who knew nothing about flying.

Sadly, for me, it came to a sudden end on the Saturday after D-Day when, instead of being with the crowd, I was spending my first night in France, hiding up in a stable.

Finally, after a miserable spell of weather and everyone wondering when the "Great Day" would be, we were called to squadron "readiness" at dawn on 6 June and heard the first reports of the massive landings on the Normandy beaches.

We waited with eager anticipation for instructions to leap into the air, but as hour followed idle hour spirits began to sag. Where was all the action? Hadn't we been prepared just for this very day?

Eventually, in late afternoon, the operational orders arrived, 129 was briefed, the "team" was listed and those selected went out to their aircraft and waited for the signal to start up.

The operation was not the search and destroy action we had expected. Instead of carrying bombs we were to escort a glider train which was to make the second big airborne drop of the day east of Caen.

As I sat in the cockpit waiting, an Avro Anson landed, taxied over to the Squadron Flights and Eddy jumped out. He ran over to the Flight marquee, spoke to one of the chaps who pointed in my direction and raced over to me and jumped up onto the wing.

"Lucky sod!" he shouted. "Things are happening and I'm missing it all!"

"Don't worry," I replied, "you'll be replacing someone soon enough."

"Well, just make sure it isn't you," he said and thumped me on the helmet as the starting flare went up.

He was to take my place within the week, bringing with him the new Mustang which I had been awaiting so impatiently.

The operation itself was a quiet one and, apart from the fact that F/Lt Hancock shot down an FW190, generally uneventful. Nevertheless it will always be remembered for the fantastic sight of all that was going on around and below. The aerial convoy of gliders and their towing aircraft was in itself an impressive sight but it was down below that the enormity of the operation became apparent.

A Channel which for years had seemed devoid of ships except for the occasional small naval convoy, had suddenly become a shipping highway. There were craft of every shape, size and function travelling in each direction, north and south, but as we approached the Bay of the Seine things became quite congested. Ships were moored off shore with landing craft plying between them and the crowded beaches; a battleship was adding its weight to the pounding of the enemy; there were barrage balloons, patrolling Spitfires and smoke everywhere.

Inland to the east of Caen the evidence of the day's earlier airdrop was evident in the broken gliders and countless parachutes which littered the flat plain, and these were suddenly added to as our own glider force began to cast off from their tugs and slip down to join the battle.

Of the battle itself we could see nothing from our height and, in fact, we were not there to enjoy the view. It was late evening when the Wing landed back at Coolham.

The following four days were a mixture of frustration and excitement as we then began the job of tackling German reinforcements making their way to the beach-head.

The weather was bad with low cloud and heavy showers but the targets were plentiful and at last we could see the effects of our work in the smoking wreckage which we left behind us.

On the day after D-Day Mark Twomey and Neil Roggenkamp got into a tangle with Fw190's, one of which was knocked down by Mark, but Roggy disappeared. Time was ultimately to reveal that having been hit from behind, he baled out and subsequently evaded capture for 10 weeks before making it back.

Saturday D+4 (10 June) was to be my last at Coolham; a bright, fine morning when I took over "readiness" duty from Les who had been up since dawn. He wasn't too pleased when the operational order to make an armed reconnaisance came through immediately after hand-over.

There were four of us on the patrol, Ruchwaldy, Humphries, myself and Thomas and as we climbed away it did not occur to me that I would never land back at Coolham and that it would be October before I flew again.

This armed reconnaisance by "A" Flight on 10 June proved to be expensive with only two of the four aircraft making it back to Coolham. Thomas at least reached the beachhead.

Having been hit by small arms fire or bomb debris I was forced to make a severe crash landing about five miles west of Vassy. The plane was written off but I stepped out (don't actually remember doing so) and started running. I was soon picked up by a local farmer who called in the help of the local schoolmaster.

Though my injuries were not serious – broken collar bone and broken nose – they were conspicuous which made it necessary to lie low for four weeks by which time the battle lines were consolidated and it was not possible to negotiate them.

It was a further five weeks (nine in all) before I managed to walk through the German lines to Patton's American forces, having twice been in refugee columns moving south.

By the time I was able to rejoin the squadron at the end of September, the Wing was at Brenzett where it had been engaged in doodlebug hunting.

My only return to Coolham was to a welcome home party at the duck farm.

Happy days!

George Frederick Pyle, F/O 172962
Yorkshire.

❒❒❒❒

How to fill in their off duty time, as well as the tales of action, is one of the memories of Flying Officer Leslie Grantham Lunn.

It is amazing how little one remembers of the everyday problems we faced in those hard and difficult times.

I do recall there was little if nothing to occupy us during off duty hours. But on one occasion George Pyle, Eddie Edwards and myself set off on a cross country walk.

We planned to walk for half an hour on each cardinal heading hoping to finish up where we started.

It was late pm when we set course following an accurate compass course. After two hours with night falling and a rising mist we were completely lost.

We had to listen for familiar noises such as cars or lorries to locate a road and it is amazing how difficult this simple exercise was.

We made it back to camp eventually but missed our supper. So much for an escape and evasion exercise!

I was the squadron imprest holder and responsible for paying squadron personnel. I collected hundreds of pounds from base accounts and, having no safe and living under canvas, it was difficult to find a safe place to keep all this cash.

Sometimes I found myself with wads of notes stuffed into my flying kit, heading out on escort duty over Germany.

At least I would have had plenty of money if shot down and if it were possible to exchange it for local currency!

129 Squadron arrived at Coolham on 6 April, in our Spitfire V's and I see from my logbook that I flew a Mustang the same day – so our new aircraft must have been waiting for us.

I remember P/O Payne being killed. They were taking off towards the lake and his throttle must have inadvertently closed as he got airborne and as he was heavily loaded with two 500lb bombs he crashed into the woods by the lake and was killed.

On and after 6 June, the invasion of Europe, times were very hectic and checking my logbook I counted nine of us killed or missing during that month of June, a third of the total number of the pilots on the squadron.

We ate in a large marquee and the food was terrible!

When we arrived at Coolham we inherited a complete Polish Wing which included Polish ground crew to service our aircraft. There was a considerable language problem between pilots and ground crew which made things difficult and consequently our aircraft serviceability dropped to a dangerous level.

We eventually received British ground crew and things improved.

A night scramble on 2 May involved taking off at dusk and heading south over the Channel in failing light and cloudy conditions, to intercept a couple of bogies.

Somehow George Pyle and I became separated which left me isolated and with a failed radio.

I found my way back to Coolham, the lake helping me to identify it. I made a couple of circuits and approached the runway over the lake and landed, quite skilfully (!) and without too many problems, to be met by Joe Leigh who said I'd been diverted to Tangmere, where George landed.

I've no idea why we were scrambled so late in the day with no night flying facilities at Coolham.

It was very frustrating on 6 June to know that the greatest military invasion ever was taking place just across the water and then not to be involved.

The squadron did get airborne later in the day but without me. I had the fit of the sulks, if I remember rightly. My turn came next day.

We were certainly heavily involved in attacking enemy transport of all types before and after D-Day.

The French were well advised to keep their heads down

as anything moving which looked suspicious was attacked, trains being one of our targets.

We always aimed for the engine which usually went up in a cloud of steam. Our object was to disrupt enemy communication as completely as possible and history has told us we were very successful.

Leslie Grantham Lunn, F/O 157825
Doncaster.

❑❑❑❑

Flight Lieutenant Arthur Charles Leigh, – nicknamed "Joe" – was on his second tour of ops with 129 Squadron when posted to Coolham and aged 24 was one of the "older" members. He had already taken part in the Battle of Britain with 64 and 72 Squadrons, completed a stint as an instructor, ferried Hurricanes from Gibraltar to Cairo that were en route for Russian pilots, been shot down over the Channel while with a Typhoon squadron and picked up by Air Sea Rescue.

As commander of "A" Flight I led its first operational mission on Mustangs on 26 April when Rocky Nelson was shot down while we were attacking Beauvais Tille aerodrome.

We had retrained on Mustangs, which none of us had flown before and done a gunnery course.

Group ordered us to attack three aerodromes one of which was Beauvais. We were just met by an avalanche of ack-ack fire from each side of the aerodrome. We had to fly through the flak and we lost one – Nelson – and two others were also hit by flak and damaged.

I was the only one to get through undamaged and decided we should fly home as I didn't want to lose any more. We only hit the one airfield, I didn't think we could go on with damaged aircraft.

We then did long range escorts and then, all of a sudden, it was decided we should do bombing as well, to work closely with the army after D-Day.

We didn't know then where the D-Day landings were going to be. Our bombing was mainly of military targets, bridges, road junctions, transport – trains, lorries, convoys on the move – interspersed with long range bomber escort duties for the Americans or British.

Briefings for the long range escorts were usually held about an hour before scheduled take off and the Wing leader would give us the target. The reading from our own base to target and wind direction would be drawn on the big map and the Wing Commander's hand would go over quite a large area and he would say: "That is where we meet the bombers".

That would last about 10–15 minutes and then we would go out to dispersal, start up, taxi out and take off.

Sometimes we would do diversionary sweeps which were away from the bombers with the idea of drawing the fighters away from the bombers.

When we were doing bombing and strafing we also had an army intelligence officer who was supposed to keep us informed on the targets we were asked to attack.

He was there mainly to be with us after D-Day because we didn't do an awful lot of escorting, mainly armed reconnaissance, and the army intelligence officer would brief us on the position of our own troops.

We had a bomb line and we weren't allowed to bomb this side of the line but we also had an area in front of the line where sometimes, if we hadn't a specific target, we kept four aircraft over an area and we could attack and bomb anything moving.

Eventually, D-Day arrived and unfortunately we were marked down as a reserve squadron. I think our air chiefs expected some fairly heavy losses and kept quite a number of Wings in reserve.

Although we got up at the unearthly hour of 3am we were on standby the whole of the day.

We knew ships were over there and landings going on and we were all eager to see what was happening. Eventually, Group agreed to us escorting gliders, some Airborne division, and the gliders landed just after Caen, but as we flew in we had our first sight of the beachhead and the sea was absolutely full of ships.

I don't think I have ever seen so many ships of different shapes and sizes all congregated together and we had to go inland a little where all these gliders were released from their towers.

They seemed to land all over the place, breaking up, wings were falling off, but a lot of blokes were coming out, and jeeps, and heading off towards wherever their official dropping zone was. I didn't see any enemy aircraft all that day but I think it was Hancock and Rigby had a quick shot at a 109 and shot it down.

They were just fortunate it happened to be their side of the squadron, I didn't actually see it. I think we were credited with one destroyed that day but all day there was very little opposition from the Luftwaffe, something we hadn't expected at all.

Going home we had another good look around the beachhead and ships, keeping fairly high because they were a little trigger happy.

This was the only trip we did on that day. After that, for the first few days, we were mainly on armed reconnaissance, doing several trips in front of the troops.

We were supposed to be posted to France so we would be in close contact with the army. A strip was to be put down for us near Caen but unfortunately the army was held up and our departure was postponed. We never did go over.

In the meantime we were bombing and strafing more or less keeping aircraft over a particular area in front of our troops.

One of our best days I led the squadron we destroyed 9 lorries and 30 damaged, it was a fairly big convoy which we bombed and then went up and down, strafing it.

It is quite different, 50 years after, but I remember going down on one lorry, strafing and firing, where troops tried to jump out of the back and get into the ditch, the sort of thing one doesn't say too much about these days.

That was more or less the pattern, flying from Coolham.

On 22 June we were told there was to be a big push from the Americans who had reached a point about half way up the Cherbourg peninsular and were stuck and couldn't seem to advance.

We were told a "big show" was going to be put on.

First of all we were told the single engine aircraft were going to strafe the front line followed by fighter/bombers and then later, light, medium and heavy bombers and the army intelligence officer said that after that lot the American army "will just walk in and take Cherbourg". He said we didn't have much to worry about as the Germans were "demoralised" and probably wouldn't put up too much resistance.

When we got there we found that wasn't quite the case. The Germans had dug in a lot of their tanks and they were using whatever guns they had on board as anti-aircraft and we went across the peninsular strafing these targets. They were spread pretty well right across the peninsular and it was really intense flak.

The squadron leader that day was the other flight commander, Hancock, and I saw him hit and go down when we were very low. He had no chance at all.

I was hit, somewhere in the oil sump and fortunately we were only a few miles from the front line and I was able to get to an inland American landing strip just south of Cherbourg.

I believe on that day the Poles lost two or three and, although I can't remember exactly, I think we lost four or five. Out of six flight commanders from the Wing I think I was the only one to get back.

I hitched a lift back in a Dakota from the American base and the first thing that happened when I landed there was one of the American fitters asking: "Do you want the clock, Buddy?" I asked why and he said: "Well, if you don't want it I'll have it." So I said: "All right I'll have it," and the Mustang's clock is still about. I gave it to the aviation museum in Norwich.

I then learned a squadron of Thunderbolts was going to bomb somewhere and I asked if I could go along to their briefing.

"Sure Buddy," they said. It was just what I imagined an American briefing to be like. "Hank, you do this . . ."

There was a big air show going on just north of them and a medium bomber came in, wheels up, pulling half the strip up and it seemed hundreds of Americans swarmed out on to the field to replace the track and get the bomber out of the way.

There was another briefing an hour or two later and they went out and just as they got ready to take off in came another bomber and pulled yards of it up again and the same process started again.

I've often wondered if they ever did bomb their target that day.

The American Dakota I hitched a lift in that evening was coming back to England with stretcher wounded.

There were a hell of a lot of other blokes obviously getting a lift back to the UK and the pilot, ready to taxi out, said: "You'd all better come up the front or we'll have a job to take off."

I was already worried that we were overloaded, there were so many stretcher cases and so many blokes standing inbetween the aisle and the middle of the aircraft. Anyway, we got out to the take off point at the end of the runway, the pilot swung it round 90 degrees and I was horrified to see behind us a glider being hooked on to the Dakota. I said to the pilot: "What are they doing here?" and he said: "They're walking wounded."

We took off, most of us standing up near the pilot and pulled this glider off and unhooked it, somewhere near Reading.

I forget the base I stayed at that night with the Dakota but got a lift back to Coolham the next day and found the squadron had already moved to Holmsley South so I had to find my way there.

About the other men serving in the squadron Wag Haw was my C/O and in addition to his flying accomplishments he was a super piano player. His boogie woogie playing, all the rage at that time, was out of this world.

Most evenings when we went out, if there was a piano anywhere around we would get Wag on it. When he sat at a piano he came to life and so did the piano.

On one occasion when we were coming down from Scotland to Hendon, just before we went to Coolham, we stopped overnight at an aerodrome not far from his home village. He was treated a little bit like a hero. Neighbours and friends of theirs put up nearly the whole of the squadron for the night.

I shared a bed with Guns Conroy, this brash, rough, tough Australian. Neither of us liked sleeping in the double bed with one another and we put the pillows down the middle so we couldn't come into contact with one another.

Wag also suffered a little from wind! I think he had a little tummy trouble. Of course, we had to maintain strict RT (radio) silence after taking off and often, when we got to about 20,000 feet, and the pressure from inside was above that on the outside, old Wag would have so much wind in him he had to let it go and over the RT would come two words, "bombs gone"!

Hancock, who was shot down in June, was a fairly quiet character who kept very much to himself.

Ruchwaldy, who took over his Flight, I had known since 1940/41 when we served together as sergeants at Hornchurch. He was born in Malaysia and had the word Malaysia across his battledress. He was a character. He also had the DFC and DFM and we were very friendly. We went around together as much as one could although he was in a different Flight. He had been rather unfortunate in his Air Force career. He had a crash and broke his back and the last I heard of him, after all the operational flying he did, he had been killed in a rather silly plane crash after the war.

The Doc, Ken Mason, was a particular friend of mine. He was a father confessor to us. Besides being a medical doctor he had to be a psychiatrist as well. Although he could drink with the best of us he did keep us in good shape physically and mentally.

When we moved to Coolham it was the first time any of us had been under canvas and I shared a tent with Doc and Ruchwaldy and Ruch and myself had thickish white sweaters and we kept those on when we went to bed.

The Doc sort of looked at us and said: "That is very unhygienic. You wear those all day, you can't wear them all night."

Anyway, in the middle of the night we heard him

scratching about, looking for his sweater, he was absolutely frozen.

Guns Conroy was quite a character, rather a tough, rough, brash Aussie, the sort one thinks of as typical. Of course, there isn't a typical Aussie any more than there is a typical anything and we had several on the squadron who were all different.

Roggenkamp was a rather superior Warrant Officer type. I don't know whether he was commissioned later or not. Most of the Aussies were Sergeants or Flight Sergeants.

Thomas was a bit unlike what we call a typical Aussie. He was not as brash and coarse as most, rather more refined.

Nick Nicholson I remember mainly because he always had cigarettes and when I ran out I was always scrounging off him and he also had a car and would take us around at night.

George Pyle and a chap named Eddy Edwards always seemed to be together and always seemed to have their white sweaters showing below their battledress top but both were good pilots.

Freddie Holmes joined us towards the end of our stay at Coolham and he had a very nice red, sporty looking MG car.

Dennis Parker went off to another Wing for a time.

Skip (Stanley) Payne got his nickname I think because he had been a steward on a liner. Unfortunately he was taking off one day very close to our dispersal – and I remember this vividly. It seemed he couldn't get full power as he came overhead. His engine seemed to lose power and he was heading for the trees and he hit the tops of the trees and went in. He had two 500lb bombs on board and my most vivid memory of that is the squadron doc, who was with us at dispersal, running towards it. I ran away from the aircraft and most of the other people ran away as well. Unfortunately, of course, Skip was killed.

Being on a fighter squadron we didn't have a lot of casualties that we actually saw. Chaps normally, when shot down, didn't come back. They were here one day and not the next, not like the bomber boys who would bring back half the crew shot to ribbons.

So that particular crash of Skip Payne's was quite a shock to all of us.

About drinking: I think I could say in the main we drank fairly heavily, but mostly beer and not very strong beer during the war. I don't remember us drinking a lot during the day. We would go out at night.

Particularly at Coolham we were on some sort of "readiness" from dawn to dusk. It didn't mean that everybody there was but it did mean we had to be at various stages of readiness from 15 to 30 minutes standby.

On 30 minutes you could be in the mess playing snooker, have a snooze, read the papers. Then you would be called to 15 minutes and would have to go to dispersal.

So we were working quite long hours but we used to go down to the local pub, the Selsey Arms, but often, if it was getting fairly late when we came off duty, we would have a few beers in the mess and go to bed.

We found one of the best hangover cures was oxygen. If we had had a heavy night out and we were probably taking off soon after dawn next morning, not feeling too bright, taxiing out we would turn the oxygen on to full and breathe it in. That helped to clear our heads.

We were due to be posted to France about six days after D-Day and it was felt that when we got out to the landing strip to be prepared for us near Caen we would find we would be short of beer. Someone came up with the idea that we utilise the drop tanks from Mustangs to cart beer from Britain to France so I was deputed to make contact with the brewery at Horsham (King & Barnes) and see if this was feasible.

I had several visits to the brewery and the brewer had a look at the drop tanks and said it seemed feasible. They would steam clean the inside of the tanks and he recommended we had mild beer as it would travel better than the bitter. I left two or three drop tanks at the brewery and maybe someone found them a few years ago and wondered what the devil they were doing there.

We were taken out of Tactical Air Force into the Air Defence of Great Britain Group and we were employed in chasing the VI buzzbombs when we left Coolham so we didn't get to the stage of carting any beer away. Of course, moving about the country I didn't go back to the brewers at all.

Regarding the food, I remember the cooks were all Poles and we didn't get on too well with the Polish menus. We asked for, and got, an English cook who tended to us.

I have a feeling there was a bit of a barrier between us and the Poles. Not language so much because most of the Poles could speak English even if some couldn't speak very much.

I suppose although we flew together as a Wing a lot of the work we were doing was in small sections and we didn't seem to mix with them as much socially as we should have done.

I think it was just a different way of life. I think they resented our being in a Polish Wing as much as we regretted not being in a British Wing.

As far as flying, generally I was quite happy flying with the Poles. The only snag was they didn't really take to bombing and strafing as we did. They preferred air fighting and in that they had no equals. They were absolutely superb. In the main we got on well and accepted one another.

My reflections 50 years on are that I do feel generally, particularly in the last 20 years, we have tended to glamourise war a little bit and it's anything but glamorous. I was often very frightened but when at Coolham I was getting on a bit, 24. I quite enjoyed bombing and strafing although sometimes taxiing out, especially if we had been given targets we knew were heavily defended, I would feel a little apprehensive. But then once we were over the target and flak coming up one didn't think too much, there was so much to do.

You had to spot the target, aim, turn round, come back again. I had been hit several times by flak and then you were a little bit scared that you wouldn't make it back.

I think you were at a low ebb probably when you went

out but having hit your target and done successfully what you set out to do there was a great feeling of elation coming home – until the next time.

An example of this was one of the chaps in my Flight was turning round and coming back from rather a lot of shows saying either his engine was rough or RT not working properly so I thought I ought to have a word with him. He started to cry and said to me: "I don't think you realise how I feel when we are taxiing out." And I told him how I felt. We found we both felt exactly the same way. We had a few more words, went and had a few beers and he was right as rain and flying again the next day.

I think towards the end of my tour I was getting a little tired and I really think I'd had enough. I think the boffins or whoever worked out the length of tours of operations had got it about right – for fighter pilots probably 200/250 hours' operational flying. For the big heavy bomber boys 30 trips might have been a bit high, 25 the Americans did. I think probably between 25/30 was about right and as much as a chap could stand. But it depended so much on the individual.

What is rather nice is the thought that after some 50 years I am going to see blokes I flew with. We placed a lot of trust in one another and I don't think I ever heard of any case of any fighters or fighter bombers that were in trouble and the other blokes, certainly his number two, were always there to help out.

I have only met two or three of them since, now I am going to meet probably 14 or 15 of the blokes I flew with, sharing good and bad times.

I sometimes see Spitfires and Mustangs on television taking off and I can hardly believe now that I actually flew, it seems so much in the past. I am so looking forward to meeting them at Coolham in June. We shall remember those we lost but shall no doubt remind one another of some of the great times we had together.

Arthur Charles Leigh, DFC, DFM, F/Lt 111975
Norfolk

❑❑❑❑

Flying Officer Johnny Byrne of the Royal Australian Air Force was 20 when he joined 129 Squadron.

We were all products of the famous Empire Air Training Schemes. R.L. Thomas trained in Canada and I completed my training in Australia.

I was only 20 years old when I joined 129 Squadron. I had my 21st birthday at Hornchurch and my 22nd back in Australia.

Between Christmas and new year, 1943-44, the squadron was withdrawn from operations and went to Peterhead in Scotland for rest.

We moved from Peterhead to Heston near London to join two Polish squadrons to form a composite Wing for the Tactical Air Force for the invasion of Europe.

Our squadron went to Llanbedr in Wales to experiment with dive-bombing a ground attack with Spitfire Mark V's.

The squadron then moved to Coolham to be equipped with P-51 Mustangs as were the two Polish squadrons. As only pilots, the adjutant and his clerk, were the only English speaking personnel there things were a bit hectic.

This airfield consisted of mesh stripping with two intersecting runways just laid on the farmer's fields through gaps in the hedges.

Because the Mustang had fairly high take off and landing speeds we practiced our first landings and take offs from Ford aerodrome, taking off on the steel mesh, flying to Ford for a few practice landings and take offs.

We also flew at 30,000 feet over the Channel towards France to testfire the guns as they were giving trouble with freezing and stopping.

A Polish pilot managed to pull the tail off his aircraft and was, of course, killed while diving over the airfield.

The Mustang would get near the sound barrier in a dive with some very real problems.

However, early in May I was posted home to Australia to join 452 RAAF Squadron in Darwin, Northern Territory.

I finished the war in Borneo and was posted home the day the atom bomb was dropped on Japan.

Joe Foster, R.J. Conroy and R.L. Thomas stayed with 129 squadron.

We all survived.

R.L. Thomas was best man at my wedding in Falkirk, Scotland in 1944 and passed away some years ago. His widow was a transport driver in the English WAAF (Women's Auxiliary Air Force). Joe Foster was best man at their wedding.

John Robert Byrne, F/O 412386 RAAF
New South Wales, Australia.

❑❑❑❑

Trapped in France being sheltered by the French Resistance during the invasion is vividly remembered by Warrant Officer Neil Roggenkamp of the Royal Australian Air Force.

When stationed at Coolham I mainly used to visit a hotel in Horsham for drinks and sometimes, for a few shillings, would hire a cake of soap and a towel for a clean-up. The squadron was under canvas at the time – no bath facilities.

I also recall sometimes walking a mile or two down the road to the village of Dial Post. On our way there we were greeted most heartily by Italian Prisoners of War tilling the fields.

In my own case I was shot down on the morning of 7 June over Normandy and after a successful parachute landing, walked south for two days, dodging the Germans on the way and eventually made contact with the French Resistance and came under their protection.

London told the Resistance to hold me there as on the night the invasion started all escape routes were closed. It was too dangerous to move any further.

I was sheltered in the village of Saint Pierre des Fleurs by Marcel Lecavelier, his wife and two young children (I was not a POW) and stayed there until the war – the fighting – came through the area, in the middle of August.

The village came under fire from both sides as the Canadian army pushed the Germans to the other side of Saint Pierre des Fleurs, where an artillery battle went on for several days.

I stayed on acting as an interpreter between the Canadian army and the village people.

Eventually I made my way back to England via Caen and Bayeux, arriving on 30 August 1944.

I was the last airman helped in that area and my stay was the longest.

As a matter of interest, in April 1987, my wife, myself and only son went back to the village after 43 years. It was a very emotional experience.

Marcel met us in Paris. We stayed with them for nine days and were treated like royalty, plenty of food and wine.

Marcel, for his work with the French Resistance, received from the Republic of France the Medaille de la France Liberee and a set of beautiful leather bound books of the history of the war autographed by Charles de Gaulle.

Neil Roggenkamp, W/O 414598 RAAF
Queensland, Australia.

❑❑❑❑

The first mention in war records of 129 in action was its participation in the raid on Dieppe in 1942, recalls Flying Officer Fred Holmes who also has reminiscences of what happened to some of the pilots posted to Coolham.

As well as putting up the money for the original complement of Spitfires the Maharajah of Mysore gave each pilot a Gundarburunda, a medal with his Coat of Arms on one side and the pilot's name on the other.

This was worn on a parachute cord hanging from the neck and beneath the tie (when worn) and every two or three months the Maharajah had his Trade Commissioner in London furnish funds for a really good party.

I was in India in 1945/46 and visited the palace at Mysore when the Viceroy was there on an official visit and I was carrying dispatches from him from New Delhi. The event was rather dull but it was nice to have been there.

While 129 was at Coolham I was a new pilot on reserve at 84 GSU. I had flown only Hurricanes and early Spitfires so I had to get time on the Mustang by ferrying them to and from squadrons and spent a few days and nights at Coolham which was a useful way to get to know everybody.

All but two of the pilots that lost their lives on the squadron, known to me, were buried abroad.

The two others, both Royal Canadian Air Force, are buried at the Brookwood cemetery, close to Pirbright, in the Canadian section.

F/Lt K.C. "Seaweed" Baker died in an accident while on leave in London at the end of September 1944. An American from New York City he had served as an instructor at one of my flying schools in Canada in 1942/43.

Some others joined 129 and also left it during that period so there was nothing like the permanence (or the feeling of it) attaching to a squadron as, for example, my father felt for his regiment in 1914/18. Two thirds of the chaps served on more than one squadron.

S/Ldr Charlton "Wag" Haw left 129 before we got to Brenzett. My mum's local paper at Rustington reported a reunion of Hurricane pilots at Tangmere in 1990 which he attended (he took Hurricanes to Russia in 1941).

F/Lt R.G. "Dutch" Kleimeyer was Australian, from Queensland, in the Royal New Zealand Air Force. He took over Joe Leigh's Flight in late 1944 and left 129 in March 1945 to join a Meteor conversion unit and therefore was one of the first jet pilots in the RAF. He now lives near me; F/Lt John Bassett, RAF, is English and lived in Montreal for many years but now retired and living elsewhere; F/O Alex Thomson, a Scottish born Canadian lawyer, lives in Quebec; F/Lt R.J. "Guns" Conroy, RAAF, returned to Australia in 1945; F/O Dickson, RNZAF, was wounded as an NCO pilot over Malta in 1942, shot down over Rouen in 1943 while with 129 and baled out, evaded capture and walked over the Pyrenees to Spain and rejoined 129. He married a girl from Shoreham in August 1945. He finished his tour with 129 and went to Hawarden as an instructor on Mustangs and is now retired.

F/O Edwards, RAF, completed his tour with 129 during which he shot down three Me109's on one show in December of 1944 over the Rhur. He later went to the Far East.

W/O Joe Foster, RAAF, returned to Australia in 1945 where he helped his father in running a snake farm, extracting venom for medicinal use.

By the end of the war W/O Tommy Hetherington, RAF, was a Pilot Officer with a DFC and left us in Norway, tour expired.

W/O Willie Jeal, RAF, was still on 129 when I left it in Norway.

F/O Leslie Lunn left us at Brenzett, tour expired. He served in the RAF post war. He earned an AFC (Auxiliary Flying Cross).

F/O Madge, squadron adjutant, left us at Brenzett. He was a county cricketer and played for Glamorgan.

F/O P.A. Nicholson, RAF, became flight commander of "B" Flight when I left 129 in Norway. He is now retired and lives in Surrey.

F/Lt Tony Osborne took over "B" Flight from Ruchwaldy and became "tour expired" when we got to Norway. He served in the RAF post war and invested in a mushroom farm near Littlehampton.

F/O Denny Parker, RAF, was posted as flight commander to 165 Squadron early in 1945. He played rugby for England in 1946/7.

W/O "Red" Redhead, RAF, was a P/O by the war's end. He left 129 in Norway, tour expired.

F/Lt Desmond Ruchwaldy, DFC, with the RAF, took over "B" Flight when Hancock was killed. He left 129 late in 1944 at the end of his third tour and was killed in a post war flying accident.

F/Lt Ian "Stalky" Strachan, an Australian in the RAF, left 129 at Andrews Field and joined 234 Squadron and later became Commanding Officer of 65 Squadron at

Peterhead, escorting aircraft to the Norwegian coast. He flew into the sea close to the end of the war.

P/O "Rocky" Nelson, RCAF, was a pre-war player with the Wembley ice hockey teams The Lions and The Monarchs. Led by Joe Leigh, he flew one of four aircraft attacking Beauvais airfield and was killed.

W/O R.L. Thomas, RAAF, left 129, tour expired, in April 1945, married a WAAF and went to Australia.

F/Lt Mark Twomey, RAF, left 129, tour expired, about March 1945. He went to Hawarden as instructor. Post war he was a commercial pilot but died of cancer in his fifties. An inveterate pipe smoker he burned the foulest of tobaccos which we suppose must have poisoned him – a lovely man, very popular on the squadron. He had a girlfriend at Coolham who ran a farm.

F/Sgt Rivett, RAF, joined 129 at Coolham and moved with it to Holmsley South. He was shot down by flak and was killed while dive-bombing V1 launching site in France. He and I travelled to 84 GSU together from Annan on 5 June 1944. 129 was short on NCO's so he went on strength before I did. He was only on 129 for about three weeks. He was 20 years old and had married on his last leave.

Jannie Bush, an Alsatian and the 129 mascot, faced quarantine if taken out of the UK so while we were at Dyce, Aberdeen, waiting to go on to Shetland and Trondheim, Mark Twomey and Guns Conroy arrived from Howarden. JB had been on 129 since it was formed almost and her owner was lost soon after she was about a year old. We had all felt concern for her future and this was the day she was to meet it.

To hoist a very large Alsatian above the rear cockpit of a Miles Master was not easy but Mark Twomey had a pup of JB's, named Monty Bush, and had brought a blanket bearing Monty's scent. This was wafted beneath JB's nose and then spread over Conroy's lap and JB settled down on it as calm as a rock.

It was a struggle to get the lid closed over the two heads but when it was done, Twomey took smartly off and we all wept with relief.

JB then became the private property of F/O "Daks" Byrne, our longtime Engineer Officer who had taken his Mustang expertise to Hawarden. He had been an RAF apprentice at Halton and was staying in the RAF post war.

I was one of many who put their name down for a job in Burma before we reached Norway.

Our postings came through after some weeks and most were still en route to India when the Japanese found out about us and surrendered . . . !

Frederick Herbert Holmes, F/O 152198
Ontario, Canada.

❐❐❐❐

One pilot who was with 129 throughout its stay at Coolham was Flying Officer Alex "Tommy" Thomson of the Royal Canadian Air Force.

I second all the complimentary references to the Poles. It was a privilege to have served with them.

As a Colonial, i.e. a Canadian, I was both amused and bemused when we joined the Poles at Heston, shortly before going to Coolham.

At one time many were singing plaintively, "An Englishman spoke in Berkeley Square" to the tune of *A Nightingale Sang . . .* It was hardly surprising that there was some muttering among the true Brits about going under foreign command in England.

The Poles, on the other hand, were less than enthusiastic when they spotted S/Ldr Wag Haw's Order of Lenin, picked up when he was with Isherwood's Hurricane Wing at Murmansk.

The Russians ran a close second to the Germans on the Polish hate list.

I suspect someone with a perverted sense of humour in Air Ministry thought it would be fun to put 129 in with the Poles.

Happily, after only a few hesitant looks at each other, we got along famously.

Of course, I remember Horsham fondly. Among other things, an artillery base there gave us the use of their facilities (Royal Engineers bomb disposal camp at Broadbridge Heath). Remember, we were in tents at Coolham and a hot bath was a luxury.

Keltz C. "Seaweed" Baker, Mike Humphries and I joined 129 at the same time.

Subsequent to Coolham I was shot down by flak over the Dutch/Belgian border in September 1944 enabling me to liberate Brussels all over again (!) before returning to the UK.

I don't think we ever had more than four RCAF members in 129 at the same time.

Seaweed Baker was a Texan in the RCAF who met an accidental death, not related to flying.

Mike Humphries force landed in Arnhem after shooting down an FW190. He was released from prison camp in 1945.

Mark Twomey, Jan Bush's nominal master, died some years ago. He also had Monty, one of Jan Bush's sons.

Alex Thomson F/O, J22313 RCAF
Quebec, Canada.

❐❐❐❐

Humour lurked never far beneath the surface however "puerile" by today's standards, according to Flight Lieutenant Ken Mason who was 129 Squadron's doctor.

My only real contribution to history was composing the squadron song at various moments.

Remembering it makes one realise how young and simple we were in those days.

I'm not certain about publishing it. What was fun 50 years ago looks pretty puerile now.

This song arose spontaneously one night when Wag Haw was at the piano. It grew unsteadily over the nights – the subject depending very largely on who was in the bar at the time.

There were probably more verses than are printed here and I'm sorry if anyone has been left out – put it down to 50 years of increasing dementia!

The title of the original music is a complete blank – the

nearest approach I've found is *The Faithful Huzzar,* a Ted Heath classic, but this is fortuitous.

There were verses about Wag Haw, Rush Ruchwaldy, Joe Leigh, Hank Hancock, Guns Conroy, Dennis Parker, Noel Green that I can remember – all equally excruciatingly bad! Wag used to play the piano to it.

Anyone reading these verses should remember that the whole world was very much less sophisticated in 1944 than it is now – but it was fun at the time!

Chorus

Oh! 129, now what is it?
It's the Mysore Squadron and bags of Bulls...,
They only fly when the weather is fine,
Those tired out boys of 129.

The squadron members

From the Russian front, the C/O came,
With the Order of Lenin and reward for same; (1)
"What does this mean, this name Mysore?
It'd better be clean, 'cos my name's Haw".

And next we come to the second in rank,
His hair is dark, his name is Hank.
He may be small, but he's not the least,
For he has been to the Middle East! (2)

Now next we must sing, if we're to be fair,
About the boy with the flaxen hair.
Let's face it, chaps (3), now where is Leigh?
He's left the room for one more p...(4)

(Ruchwaldy's verse – how did it start?
I just can't think, but here's a part):
"Though most of my time is spent in bed,
My home of rest is Peterhead." (5)

From Australia, there came a shout:
"Oh dear, oh dear! get in, get out" (6)
His RT voice scares all the Huns,
You'll hear them say: "Achtung! It's Guns!"

"There is a man called Skipper Payne (7)
who used to sail the Spanish Main,
We see him, now, in a Supermarine
Up in the blue, not on the green."

Said little Denny Parker to Little Noel Green:
"You haven't been around, where have you been?"
Said little Noel Green: "I've been eating my pud,
And you must know – If it's Green's it's good." (8)

Now, who d'you think writes all this rot?
It's got to be the Squadron doc.
"If you want me to see any sore or scar,
Tell the orderly: I'll be in the bar."

(1) The Order of Lenin carried a stipend of approximately £5 per year.

(2) At this time, the greatest "line shoot" was to have been in action in North Africa.

(3) Joe's constant admonition when things were getting out of hand.

(4) And he had a notoriously weak bladder. It's an interesting social comment that, in those days, we would obliterate such a rude word if WAAFs were in the bar.

(5) Ruch had an immense capacity for sleeping in in the morning. He had done at least two tours at Peterhead before 129 moved there.

(6) Two of Guns Conroy's favourite expressions.

(7) Skip Payne, I think, got his nickname because he had been a steward on a liner.

(8) The advertising slogan at the time for Green's Sponge Pudding.

The only clear memory I have of Coolham (other than Joe and Ruch coming back to the tent to tell me about D-Day) demonstrates some of the high-powered medicine I was prone to at the time.

Some of the boys, led, I think, by John Bassett, cornered me and said: "Doc, you've got to do something about the tea in the mess tent – it's quite horrible."

So, off I go to the Sergeant Cook – who was, of course, Polish – and asked him if the tea had changed.

"No," he said. "It's the same tea we always get from the NAAFI." So I asked if, perhaps, he was making it differently?

"Not really," he said. "We have a little trouble because we've run out of sugar. But everyone knows you have to put something in the Britishers' tea so I put salt in instead!"

Back to the present day I would like to see Guns Conroy's hand which I stitched up while he held the room steady with his other arm – it was a beautiful job at the time, but not done at Coolham.

J.K. Mason, F/Lt 148693
Professor (Emeritus) of Forensic Medicine
University of Edinburgh, CBE MD LLD FRCPath DMJ

❑❑❑❑

I am glad to say that I was stationed at Coolham with 129 Squadron and what a hectic few months it was, writes Leading Aircraftman Harold Wilson, a fitter.

We thought we were due for overseas at first until we settled and saw the laid out metal runways and dispersal points.

We soon settled down and eight to a bell tent was just fine.

We were kept pretty busy all the time escorting and loading up for bombing strikes.

I did the DI (Duty Inspection) in a morning – petrol, oil coolant checks and running the engines up. There was nothing like the roar of the merlins in the Mustang!

My memory fails me a lot regarding the things we did, also regarding the names of all of the lads.

Still my thoughts are there.

Harold Wilson, LAC 1048581
Deeside, Clwyd

❑❑❑❑

Flight Lieutenant Noel Sturt Green was killed on 17 June 1944 while stationed at Coolham with 129. He was shot down over France at the age of 21.

His family owned the Green's cake company and came from Worthing.

He was sent to America to be trained and when he was an aviation cadet at the Advanced Flying School, Craig Field, Selma, Alabama, USA, he wrote this letter to his sister Sheila on 29 January 1942, which she kept.

I am writing this just before I am due to go night flying. I like night flying very much and it may surprise you to hear that I am trying to be a night-fighter. This camp is a fine place and huge landing runways. We do all kinds of flying, formation and gunnery practice. I sincerely hope to be back in England before the end of March but I can hardly tell how long we shall have to wait in Canada for a boat. Anyway, the quicker I get back to England the better.

I have had quite a nice lot of letters from Daphne and she seems very well. As I expect you know she is stationed up north. When I get back to England I shall send a telegram to Ma and tell her I am back in England and then find out if Daphne is still on the same station. Then I shall go to Daphne and fetch her and bring her down to Worthing with me. You see she will get seven days leave at the same time as me. On passionate grounds, if you understand me.

I have told Ma some of this but naturally not all of it. I expect you know by now, I am crazy about her and I think she is about me so things should turn out pretty well, especially if we can get leave together.

I bet she looks fine in her officer's uniform. I hope I get a commission but I am not sure whether I will or not. She sent me a photograph of herself in WAAF uniform and she looked fine in that.

I am beginning to collect things to bring home to you but I don't know what to get for you. I expect I will find something as a present from America.

The weather here is quite nice, not very cold and sun shining most of the day. It is really lovely up in the sky with not a thing to worry about. I have got about 150 hours' flying now and by the time this course is finished I shall have 200 hours.

As I expect you know, we get given RAF and American Wings and also an honorary commission in the US Army Air Corps.

I hope you are doing well in school, especially at games. I never did a stroke of work at Blundells but had a good time in the sport line. I hope you are working hard for your School Cert.

I have met a lad who lives in Burgess Hill and we are great pals. When we get back I shall take Daphne over to Brighton and meet him and his girl there and probably paint the town red. Oh boy, I am longing to get back for so many reasons.

It will be strange when I get back to drive on the left hand side and see more white people than black. In Selma over 75 per cent of the people are niggers.

The food is good over here. Everything that you want. Oranges, chocolates, sweets, butter, sugar, meat, cigarettes, in fact, everything under the sun.

Since I have been over here I have been to these States – if you can find them on the map you can probably see them better – Washington, North Carolina, South Carolina, Georgia, Alabama, Florida and Ohio. I may go through a few more by the time I am finished over here. The nearest big town to Selma is Montgomery. I haven't been there yet but hope to go on my next leave out of camp.

I am bringing back one coin of each value in America when I come back. They are, with their approximate English value: the dollar (a note) = 4/6d; silver coins of 50 cents = 2/6d; a quarter = 25 cents = 1/-; a dime = 10 cents = 6d; nickel = 5 cents = 2d, and 1 cent (brass) = ½d.

So you will be able to see what kind of money I had to use.

Well, old lass, I must close now and do some night flying. Keep smiling, not long now before we will be together again.

Tons of love and kisses from your loving brother, Noel.

PS: I will try and bring some stamps back for you if possible.

Noel Sturt Green, F/Lt 122119

❐❐❐❐

Pilot Officer Douglas Haig Nelson, Royal Canadian Air Force, was a member of 129 Squadron killed in action on 26 April 1944.

This is his last letter to his sister, Jean Hurndall, dated 20 March 1944, a copy of which was given to Paul Hamlin when he met her in Canada in August 1995.

How goes everything, Jeannie, hear you have been ill but trust all is well by now.

Have heard from Mom, am pleased she has at last arrived in New York. The two of them will have a wonderful visit. Vera will now be lonesome for a while.

Well, at long last my commission is through, back dated to January. The advantages are already noticeable, just a dinner in the mess of course, but will soon settle in. Was quite an old soldier in the Sgts mess.

Should be at my new station tomorrow, thank goodness, our rest is over. The next one will be when I finish a tour and will last for six months or so, and then with all the luck (who knows) possible maybe a month's leave at home. That will not be for a year though so am being a little too anxious.

Have had an aspirin poisoning again, will not take them again unless a day's rest is required! Very handy.

Had to force land last Friday but got down with no harm to self or machine. Still waiting to get up again possibly this afternoon or tomorrow morning. No one seems to be in a hurry but myself and I am getting fed up waiting.

Spring is at last on the way, will be nice to have some nice weather for a change. Hope to have a bird's eye view of enemy country soon. Wish me luck.

Will away for now Jean – pray for me and wish me luck.

Best regards to Harold and all. Heaps of Love, Doug.

Douglas Haig Nelson, P/P 54296, RCAF

❐❐❐❐

Flying Officer Mike Humphries joined the Royal Canadian Air Force in Toronto in 1941, his family having emigrated there from England in 1927. After two years of flying Hurricanes off the east coast of Canada and several months at an Officer Training Unit in Shropshire he was finally posted to 129 (Mysore) at Peterhead, Scotland on 10 March 1944.

Mike Humphries force landed in the Arnhem area of

Holland on 25 September 1944, with engine failure after shooting down two Fw190's and while chasing a third.

He was taken prisoner of war by German Panzer Grenadiers, taken to a POW camp and arrived back in London about nine months later.

Mike, a professor of psychology and dedicated conservationist, was killed piloting his own Cessna 172 in British Columbia, Canada, on 31 August 1995 while carrying out video work for Lighthawk, Conservation Airforce, over Mt Curran, Vancouver Island. He was 72.

My wish had been granted. I was to go to an RAF squadron rather than an RCAF one, a choice I had made at OTU because I had heard that the Canadian squadrons were committed to clean buttons, polished shoes and parade square bashing. That was something, as it turned out, 129 could never be accused of being.

My introduction to the squadron was made at the bar in Peterhead where I was greeted by S/Ldr Wag Haw and a gaggle of 129's pilots.

In very short order it became clear that no one was much interested in my flying experience; the only concern seemed to be how rapidly I could down a pint of ale.

Apparently they were much in need of a rapid quaffer to bolster their beer drinking team. The form was explained and demonstrated and I was put to the test.

The beer contests were much the same wherever we went. Two teams would face off against one another across a long table, or a series of smaller ones if they were the only ones available, long enough to accommodate 10 to 12 pilots on each team.

Mugs were filled and with appropriate solemnity the word "GO" was pronounced. The first pilots of each team would then down their pints as rapidly as possible and invert their mugs on the table with a crash, as a signal for the next in line to raise and down theirs. And so on down the line.

The losing team usually had to provide refills for the winners, who would then challenge any other group still standing. As with many relay races the anchor man was considered critical. He had to assess the chance of winning and if the odds were poor he would have to make special efforts to inhale his potion in an unbelievably short span of time.

I had developed a fine skill while in Gander, Newfoundland. I could open my throat and simply pour the beer straight into my stomach. My beer downing ability instantly placed me in a position of great responsibility and I was appointed to the Mysore team immediately and the gauntlet was dropped on the members of another squadron.

Several hours later I must have been shown to a cot in the barracks but I have no memory of it.

I do remember waking in the morning with a horrible headache.

I was informed that the squadron would be flying the Spitfire Vb, a version that had been designed for low level operations. We were posted to Heston and over the next few days we were busy packing up. The evenings were spent hangar flying and sharpening our quaffing skills.

I was also introduced to a form of parlour rugby which involved lining up against an opposing team in an attempt to push them up against an opposite wall or into a fireplace or out some door.

There was a form of obstacle race entailing lining up sofas, stuffed chairs and other assorted impediments around the main mess lounge.

Individuals would then be timed as they proceeded from start to finish, diving into chairs and sofas, leaping coffee tables and speeding from a pint at the start to a pint at the finish line.

I suspect in the history of human conflict more injuries were sustained by the RAF in this harmless pursuit than by all other non-violent means of relaxation. All too often the squadron doctor was involved in the fracas so that by the end of the evening his abilities to repair damage were somewhat impaired.

Some of the pilots were designated to fly to Heston, the rest of us were to go by train and I made sure that Jannie Bush (JB), a German Shepherd dog and the squadron's mascot, was comfortably ensconced in the baggage car.

It was here that I first flew the Spitfire Vb but the most memorable experience at Heston was my introduction to the Blue Ball Club (the BBC) by one of the 129 crew. Ostensibly a private affair it was located on the left hand side of Blue Ball Yard, just off St James Street, close to Piccadilly in London.

It opened around noon, was usually quiet in the afternoon, filling up about 4.30pm until by 6pm you were lucky to get close enough to the bar to get another drink.

Many of the people there were in some service but often as not they were in mufti. We suspected that some were involved in some sort of clandestine operation.

There were a number of WRNS and WAAFs who gravitated to the club.

Frequently, a group would decide to find a Chinese restaurant somewhere and amazingly, in London at that time, you could still get very good Chinese meals for reasonable prices. Then it was on to some dance or back to the club.

The key to the BBC was Bobby. She had an amazing elan and energy. She seemed to know all the many people who passed through her doors and welcomed everyone.

She was possibly 30ish and seemed able to detect when someone needed cheering up and when they needed to be left to think in peace.

I recall several trips with her, later on, when we would drive in a small van to the East India Docks to pick up liquor and cigarettes. Usually these forays were late at night for reasons one could only guess.

Bobby had a charming, many-roomed apartment over Prunier's Restaurant at the bottom end of St James's. There was a large living room, several bedrooms, a kitchen and ample storage.

In the months ahead I and others in the squadron were to come to the apartment fairly frequently, partly to have a bath and partly to pick up a change in clothes.

One of the delights of Bobby's apartment was the dumb waiter that connected to the restaurant below. We would phone down, place an order and in due course a bell would ring and we would pull up our order, complete with bill. We would then send our payment back down on the dumb waiter with the dirty dishes.

It was at the BBC that I met and fell in love with E..... She was the barmaid at the club, attractive, blond and at 22 or so, quite mature in the eyes of a young, 20 year old pilot with very limited experience.

Our affair, which started after the squadron moved to Coolham, was not to last very long. I suspect I was much too immature to appeal to E for long. In any event, Seaweed Baker, one of my tent-mates, replaced me in her affections, which had a devastating affect on me, for a week or so, just about the time we moved from Coolham to Ford.

I continued to frequent the BBC even after the break-up with E and I, with others, continued to use Bobby's flat as a way station for clothes and baths.

Eventually, after a brief spell in Wales, we flew in to Coolham on 6 April 1944.

We circled the field to inspect the runways and to get set up for landing. What I saw from 2,000 feet or so was relatively lush field surrounded by trees, a small village, a manor house with pond and stream on the north side and the dim outlines of the Sommerfeld tracking.

The tents that were to provide housing, mess halls, maintenance and dispersal areas could be seen nestled under the trees. A very peaceful, pastoral place compared to the concrete, steel and hard edge surfaces usually associated with airfields.

The meadow became our playing field for rugger and football in the weeks to come.

That evening we walked up the road to the Selsey Arms pub where I fear our crowding presence must have made some of the locals a little uncomfortable, but nothing was ever said to us. We played darts and quaffed copious quantities of bitters and a local ale.

The following day, 7 April, I had my first flight in the Mustang. As a plane to fly I never enjoyed it as much as I had the Spit. Not sure quite why. But it was stable and seemed not to have any serious tricks.

For the next week or so we practiced aerobatics, formation flying and cross country work to become familiar with the area immediately around the airfield. By 25 April the Wing was operational and ready for what we knew to be around the corner – the Second Front.

Coolham Airfield was designated an Advance Landing Ground, both a staging post and a practice site for what was to be a highly mobile fighter wing, capable of operating from relatively simple airstrips, completely self-sufficient. Being highly mobile and independent of fixed bases, in theory we could move forward with the ground troops to provide air cover and ground support wherever they advanced.

Most of April was taken up with the process of becoming familiar with our new fighter, the Packard Merlin powered Mustang P-51.

Some of the more experienced pilots did begin operations toward the end of April, in fact we lost P/O Rocky Nelson on a raid over France on 26 April. He was reported to have bailed out but we later learned that he had been killed. On that flight Guns Conroy and Byrne were also hit by flak but made it back safely.

During April we also got to know the pilots of the two Polish squadrons. They were generally excellent and enthusiastic characters, somewhat impulsive at times, but who wasn't. Somehow they became dubbed "Those Crazy Poles" for reasons no longer clear to me.

On 1 May the designated pilots from all three squadrons were assembled in the Operations Tent, where we were briefed on the mission, to escort a force of Boston bombers to Charleroi in Belgium, just south of Brussels. As in most briefings, we were shown maps of the area, the proposed track of the bomber force and the intercept region where we were to make visual contact with the bombers.

The call letters were assigned and the radio frequencies in use were provided. Areas of expected heavy flak were located for us on the main map and the recommended courses to avoid them were agreed upon. Finally, the take off time was specified and the procedure for getting the Wing formed up was defined.

We went to Squadron Dispersal to pick up our gear and to clear our pockets of any items that were considered a security risk in the event we went down and were captured.

Finally we headed out to the designated aircraft. I found DV-M located behind some trees where the ground crew assured me the aircraft was in good shape and helped me to settle into the cockpit.

A short time later a flare appeared over the Control Tower and we started up our engines and taxied for the take off runway.

In the next 10 minutes or so, 36 Mustangs roared down the runway and in due course formed up as a Wing prior to departing across the English Channel.

For me it was an uneventful mission as we saw no enemy aircraft and the flak, although quite heavy around the bombers, was no threat to us. But I do remember my first sight of France near Calais. Strange feeling. I had to force myself to realise that while we were flying over enemy territory, the country itself, although occupied by Germans, was friendly.

There was a small frisson of tension as we searched the sky for 109's and 190's, after all, our task was to protect the bomber force.

Although flying in fairly close "finger formation" I tried to keep track of our position and to keep a mental picture of the route home in the event that I was to become separated from the squadron and had to return alone.

Eventually, the bombers dropped their loads and wheeled for home. We maintained our position, criss-crossing above them until we crossed the French coast. We then broke off, let down across the water and returned to Coolham.

The squadrons separated, strung themselves out in two-plane sections, and landed as quickly as possible.

Usually we knew the night before who was scheduled to fly the next day. This governed to some extent how much pub crawling you did the night before a flight.

During the month of May at least one operation was

carried out each day, weather permitting. Many of the operations that month were relatively short range affairs aimed at sites in France and Belgium. Our targets tended to be railway marshalling yards, railway bridges, surface transport and what were then called "Noball Targets", a code name for what we were later to call "Divers, Buzzbombs and Doodlebugs", or simply "V1's". Other flights were into Germany in support of the daylight raids being made by US Flying Fortresses to towns such as Hamburg, Brunswick, Magdeburg and Strasbourg.

A typical long range escort mission began with the usual process of forming up the Wing somewhere between Coolham and Manston, a climb to assigned altitude, usually above 25,000 feet or so, a rendezvous with the bomber stream as it crossed over France, Belgium or Holland, or we would catch up to the stream later in their mission, over the target, or on their withdrawal from Germany.

Along with other long range fighters of the RAF and the USAF the object was to provide continuous fighter cover for the bomber stream, to deter attacks by 109's and 190's. Sometimes you could literally fly along the bomber stream, as it stretched out over many miles, which made navigation easy.

Sometimes you could locate the stream far ahead simply by locating the bursts of flak. Quite often, particularly when we were serving as close escort, the flak was aimed at us. Without warning a great black cloud would suddenly appear near you, sometimes you could smell the burned explosive as you flew through it or literally feel the shock wave if the explosion was close enough.

I think most of us picked up some shrapnel that month but we didn't lose any aircraft.

The memorable aspect of those flights was their duration – over four hours on some occasions. Quite often we would see no enemy activity, other than the flak bursts, and more often than not we would return not having fired a shot.

Sometime that month we had a visit from an Air Vice Marshal (RAF) from HQ, a blustering fellow with red face and blond eyelashes and eyebrows.

His mission was to check on our operational readiness and to stir our morale for the task ahead – the Second Front, or D-Day as it was to be dubbed later.

However, in closing his final address to us, almost as an afterthought, he commented on the frequency with which we were returning from raids over Germany without having fired a shot. He almost ordered us to ensure that we make each flight count; if we had not used our guns, we were to drop down before leaving Germany and strafe anything moving including "Hausfraus hanging up laundry".

I think we were all dumbfounded; we were silent, as I recall, until he had departed. Later, we were to discuss his order, if that is what it really was, in amazement.

Not only would it have jeopardized the main mission, protection of the bomber force, which we could hardly do at ground level, but, more importantly, for most of us, it seemed contrary to what we took to be the moral basis of our role.

To purposely attack unarmed civilians was not part of our concept of our role. We were aware that bombers did it, but fighters!

Our targets were military, only, or that was how we saw ourselves. Perhaps not very critically, I now admit. But the Air Vice Marshal's attitude was not then part of our perception, if it ever was for most fighter pilots.

As far as I know, at least in 129, the order to strafe civilians was simply ignored and we would continue to return without firing our guns unless we encountered military targets.

My last flight in a Spitfire happened on 16 May, a ferry flight to Hamble on the south coast. On the way I took the plane up to 10,000 feet and did a series of joyous aerobatics before landing at my destination. It was too late to catch a train back to Horsham so I stayed the night on a live-aboard sailboat owned by one of the repair unit's pilots.

I dimly remember making an uncertain journey over many planks between a number of small vessels before we arrived safely.

The next day I caught a train that deposited me in Horsham, just in time for a squadron bash in the largest local pub.

One man I particularly recall at Coolham is Blake (Tex) Bailey, a technical specialist with North American Aviation, the manufacturer of the Mustang, who was placed with 133 Wing to advise on maintenance and repair.

Tex Bailey was unlike any other Texan I had met up to that time. He was quiet, unassuming and gentle and truly curious about things. Unlike many from North America he was never discourteous or derisive in his observations about the technology, or the customs, we found somewhat strange in wartime England. He seemed to get along famously with the Polish ground crew and technicians.

Tex and I talked for hours about books, and life in the southern United States about which I was puzzled.

I was having trouble understanding the usual southerner's attitudes to the negroes, at least the attitudes I had picked up from other southern pilots who had come north to join the RCAF in 1941.

Tex was different – a different sensibility, more like the northerners I had met.

We also visited the Blue Ball Club when in London and I made use of his plush hotel rooms from time to time.

I spent many hours with Tex at his tent "office" near the intersection of the two runways at Coolham. I'd wander over to look at engine parts and systems he was analysing. I'd go with him when he was working with the ground crew while diagnosing particularly nasty system failures.

His patience was immense. And in the course of all these "seminars" I learned a lot about the Mustang III and a good deal more about how to get along with machines and people. He was a great teacher.

On another level I have fond memories of Jannie Bush, our squadron mascot, an unusual German Shepherd dog. I had accompanied her south from Peterhead in March and I was to spend much of the summer with her walking and exploring the countryside.

She usually slept in someone else's tent but would be around early in the morning to see if anyone wanted to play or go for a walk, checking each tent for likely prospects.

I remember May and parts of June with pleasure. Coolham was a great place to be that spring. Once the morning flight had departed a strange silence fell on the airfield. Now and then an engine would fire up as it was being tested and then it would shut down and the silence would return.

I was amazed at the variety of birds and that they seemed little disturbed by all the noise we created for brief periods of time. And there were many balmy days when the warm spring sunshine brought out the flowers and invited us to remove sweaters and shirts.

Soon after our arrival I discovered the little lake behind our dispersal tent and I would wander along it with JB looking for frogs and signs of fish.

Someone at the Selsey Arms pub suggested there were pike in the pond so I began to think of tackle and how I might catch one.

I think it was Joe Leigh, my flight commander, who suggested that it might be proper if I asked permission to fish the pond.

So one day, with JB in tow, I wandered up to the house above the pond and duly obtained the needed approval. I fashioned a lure from a tin can and a hook obtained from one of the chaps at the pub and with some black, heavy thread used to wrap electrical wires, I attacked the pond with a crude throw line.

JB would sit beside me obviously puzzled but clearly content to be exploring new territory. We would then follow the little creek that flowed east and somewhat south out of the pond. She enjoyed swimming after the ducks that inhabited the wetlands and I suspect she got closer to a "kill" than I ever did with my fishing efforts.

But it was a lovely peaceful way to spend those last lovely days of spring, days when I was not flying across the channel to harass the Germans and inconvenience the French.

JB would often go with us on our pub crawling expeditions. In most cases she was allowed to sit with us in the pub or, on pleasant evenings, out on the lawns, beside the attractive little rivers in southern Sussex.

She would sit up in the jeep, with her head stuck out into the slipstream, as most dogs will do. JB had a nobility about her, a calm lightness of being that few humans can equal. A great companion in most difficult times.

Throughout the spring of 1944 in southern England tensions had been rising. We knew the Second Front was about to be opened but we didn't know when.

Coolham was visited by senior officers from the Second Tactical Air Force to boost morale and generally urge us on. Gliders landed, paratroops practiced by dropping in on us and everywhere you drove you encountered masses of vehicles and troops sitting around with nothing to do.

For ourselves, we had trained and practiced every skill related to our work until it had become routine and on raids into France had put into effect what were otherwise theoretical abstractions.

June began with bad weather – low, scattered and broken cloud under heavy overcast, with fairly high winds and rain.

On 2 June I was part of a four-plane Ranger to the Namur–Verdun area. We attacked a dozen barges in the canal and left some of the barges burning.

On the 5th I flew down to Ford, an aerodrome near Arundel in Sussex, to have a brief visit with some of the old 127 mates who had become part of Johnnie Johnson's new Canadian wing flying Spitfire IXs.

And then it happened at last. On the 6th, the Second Front finally got under way.

Our squadron carried out a Ramrod mission near Caen to cover the airborne forces landing near the beach. Hancock and Rigby shared an Fw190. On the 7th the squadron conducted an armed reconnaissance of the Argentan area. We counted numerous convoys of trucks and armoured vehicles moving toward the beach. We attacked the convoys and a train with bombs and machinegun fire.

Roggencamp went down, hit by flak, and Mark Twomey destroyed an Fw190. I picked up some shrapnel in my tailplane. Later in the day I made a second reconnaissance in the Argentan area. Again we dropped bombs on columns of motor transport and tanks and proceeded to attack a very long column of transport, including horse drawn trucks. Very accurate flak. I picked up some more shrapnel but nothing serious.

And so it went. Up early for the dawn flight, several flights a day, never more than 2½ hours. Often the last flight of the day did not get down until dark. The bulk of our work was at very low level and most of our attacks were against shipping, troop and transport convoys, flak positions, tanks and troop concentrations.

During the month I managed to pick up flak damage on very nearly every trip, one of which was a serious glycol leak that forced me down at Ford.

We lost many pilots that month, either killed in action, taken prisoner behind the lines and others who simply went missing to turn up months later having escaped through Spain to return to England: Hancock, Roggenkamp, Thomas, Pyle, Rigby. Several made safe landings either on the beach in France or at a drome back in England.

Payne was killed on take off from Coolham after his engine failed to develop power and dropped him into the trees at the end of the runway. Rivett dived into a wood east of Paris early in July.

On the 20th Joe Leigh sent me to Honeley to instruct for a few days, after which I returned to the squadron at Holmsley South where the squadron had been moved to await orders to move to France, orders that never came: the buzzbomb raids on London had begun.

Regarding Skip Payne, I never knew him very well. I remember playing darts with him at the Selsey Arms, but not much more, until the day he crashed.

I was standing outside our dispersal tent that day, 17 June, watching "B" Flight take off for a Ranger in the Caen area. I just happened to be watching as Skip's plane left the airstrip and then started to sink toward the trees beyond the pond.

I saw him hit the trees, heard the sounds and watched as the plane sank out of sight near the Manor. I remember running as fast as I could, up the footpath beside the pond, around the house and along the hedgerow toward the crash site.

As I entered the field I could see the plane, nose down, against the hedge. As I ran the remaining distance to the plane I could make out Payne's head resting forward on the instrument panel, motionless. Just as I was reaching a point about 200 feet from the aircraft, there was a small explosion from the engine compartment and flames quickly engulfed the forward half of the plane obscuring the cockpit entirely.

The heat was intense; it drove me back, leaving me in horror as I watched helplessly. Others began to arrive, also powerless to help.

As it was, I flew a Ranger later that day to the beach, without incident, had some beer in the mess tent that night and then lay awake for hours before sleep finally rescued me.

We tried to analyse what must have happened and concluded that Skip either experienced an engine failure, or he suffered the gradual loss of power as the throttle slowly moved back from the full power position.

We had learned that the friction control had to be really tight, otherwise the throttle would indeed creep back. Perhaps that was why we usually flew with our left hand on the throttle all the time.

In Skip's case he would have removed his hand from the throttle to raise his undercarriage, just about the time he sank into the trees.

We also concluded that from the position I had seen him in, motionless against the instrument panel, he was probably unconscious from a head injury at the time the flames erupted.

Mike Humphries, F/O J-3609, RCAF

1944, P/O Mike Humphries in civilian guise, RAF escape photograph.

Photo: R.J. Conroy

1995, Mike Humphries, British Columbia, Canada. *Photo: P. Hamlin*

129 (MYSORE) SQUADRON
RAF Peterhead, January 1944

Back row, from left: Sgt Bois, Sgt Lane, Sgt Lewis, Sgt Hillier, LAC unknown, LAC Hodgson, LAC unknown.

Centre: F/Lt I.D. S Strachan, W/O N. Roggenkamp, W/O Rigby, F/O S. Payne, W/O G. Cliff, W/O A.J. Foster, W/O J.R. Byrne, F/Lt J.K. Mason, F/O M. Twomey, W/O R.L. Thomas, P/O D.H. Nelson, P/O E.W. Edwards, P/O G.F. Pyle, Sgt Fransman.

Seated: F/Lt N.S. Green, F/Lt J.P. Bassett, F/O D.C. Parker, F/Lt A.J. Hancock, S/Ldr C. Haw, F/Lt A.C. Leigh, F/Lt D.F. Ruchwaldy, F/Lt R.J. Conroy, F/O D.C. Byrne and JB – Jannie Bush, squadron mascot.

Photo:R.J. Conroy

P-51b Mustang, serial No. FB 108, prior to delivery to 129 (Mysore) squadron. On 10 June 1944 P/O Pyle crash landed this aircraft in enemy territory. *Photo: Brenzett Aeronautical Museum Trust, Kent*

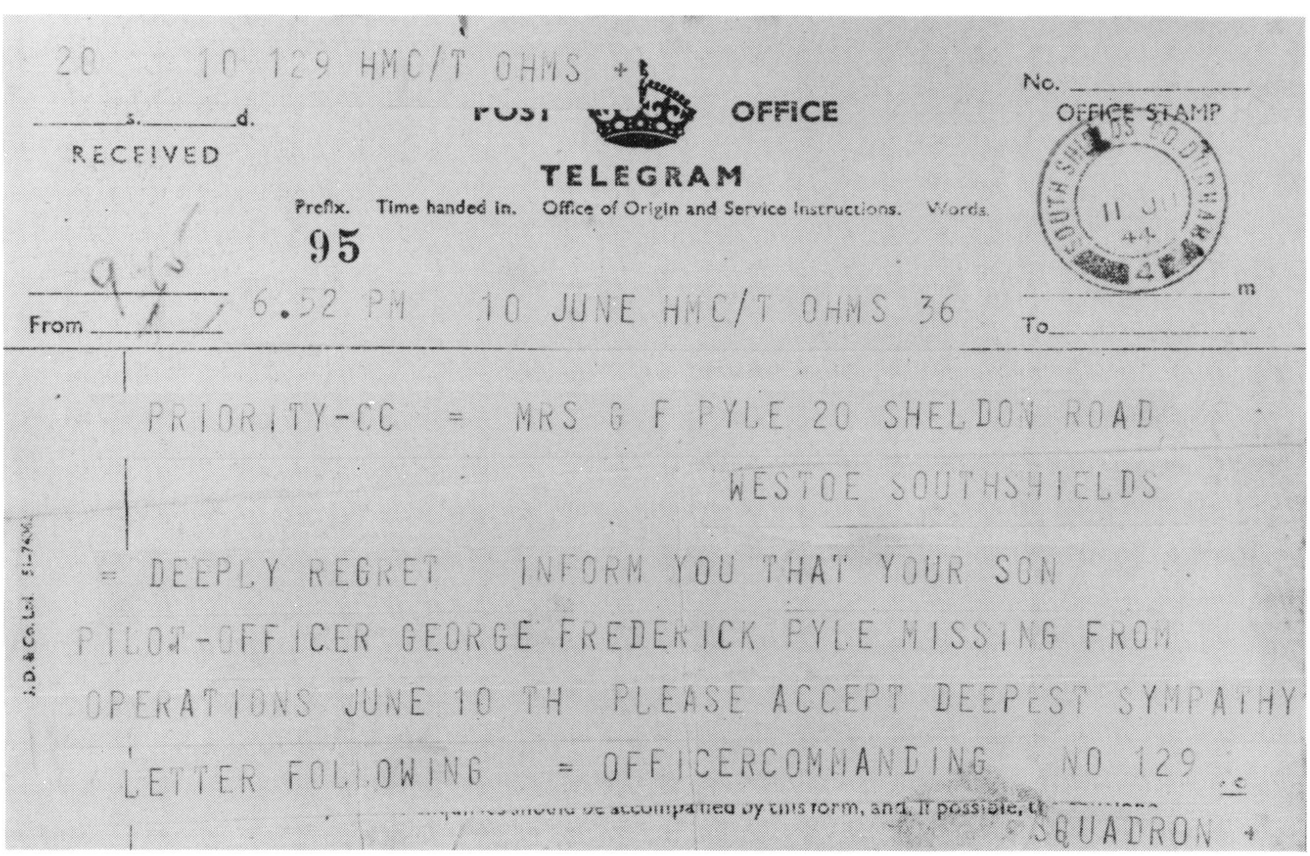

20 10 129 HMC/T OHMS +

No.

POST OFFICE

s. d.

OFFICE STAMP

RECEIVED

TELEGRAM

SOUTH SHIELDS CO DURHAM 11 JU 44

Prefix. Time handed in. Office of Origin and Service Instructions. Words.

95

From 6.52 PM 10 JUNE HMC/T OHMS 36

To

PRIORITY-CC = MRS G F PYLE 20 SHELDON ROAD
WESTOE SOUTHSHIELDS

= DEEPLY REGRET INFORM YOU THAT YOUR SON
PILOT-OFFICER GEORGE FREDERICK PYLE MISSING FROM
OPERATIONS JUNE 10 TH PLEASE ACCEPT DEEPEST SYMPATHY
LETTER FOLLOWING = OFFICERCOMMANDING NO 129
SQUADRON +

...should be accompanied by this form, and, if possible, th...

Post Office telegram informing next of kin that P/O Pyle was missing from operations. *G.F. Pyle*

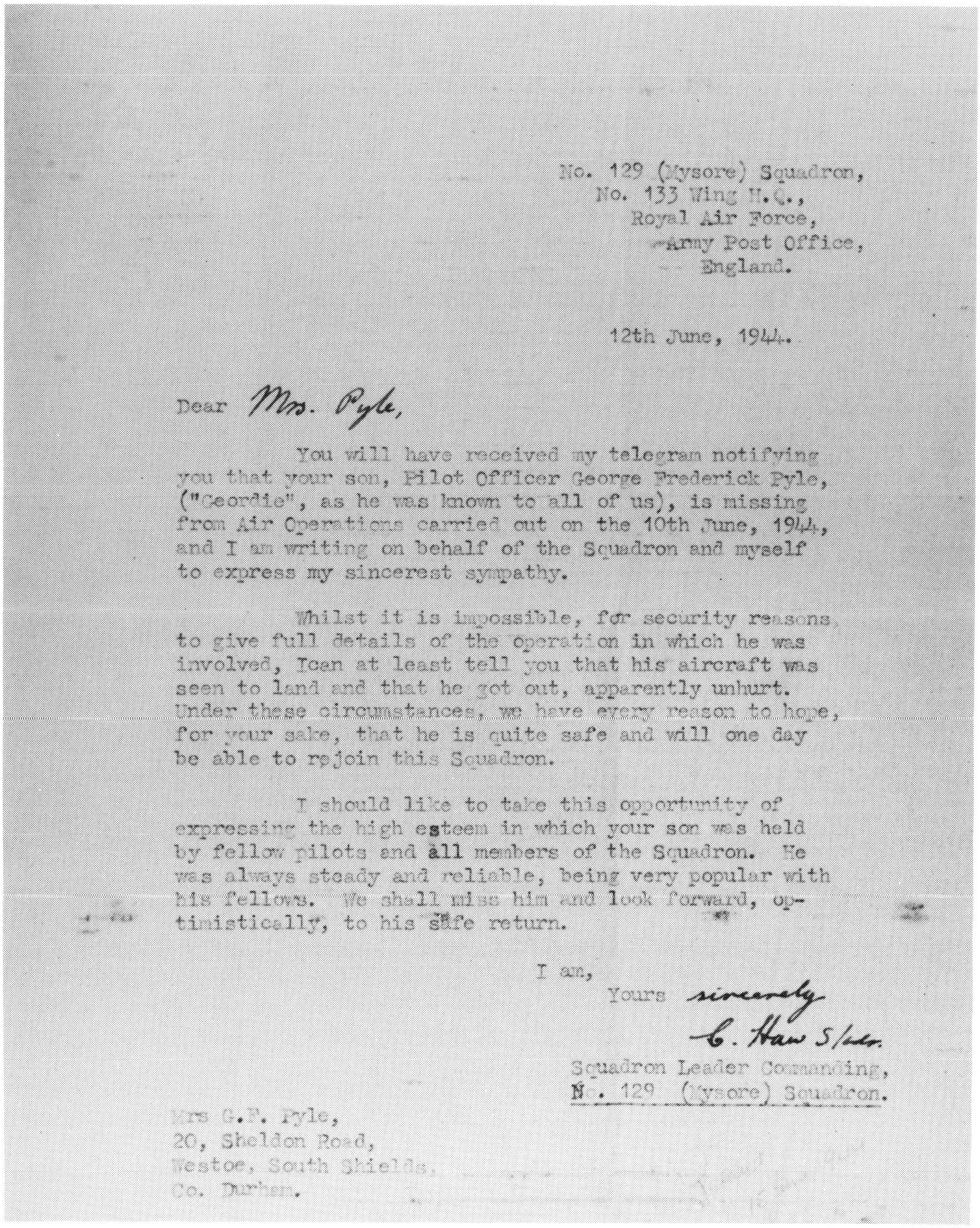

No. 129 (Mysore) Squadron,
No. 133 Wing H.Q.,
Royal Air Force,
Army Post Office,
England.

12th June, 1944.

Dear Mrs. Pyle,

You will have received my telegram notifying you that your son, Pilot Officer George Frederick Pyle, ("Geordie", as he was known to all of us), is missing from Air Operations carried out on the 10th June, 1944, and I am writing on behalf of the Squadron and myself to express my sincerest sympathy.

Whilst it is impossible, for security reasons to give full details of the operation in which he was involved, Ican at least tell you that his aircraft was seen to land and that he got out, apparently unhurt. Under these circumstances, we have every reason to hope, for your sake, that he is quite safe and will one day be able to rejoin this Squadron.

I should like to take this opportunity of expressing the high esteem in which your son was held by fellow pilots and all members of the Squadron. He was always steady and reliable, being very popular with his fellows. We shall miss him and look forward, optimistically, to his safe return.

I am,
Yours sincerely
C. Haw S/Ldr.
Squadron Leader Commanding,
No. 129 (Mysore) Squadron.

Mrs G.F. Pyle,
20, Sheldon Road,
Westoe, South Shields,
Co. Durham.

Personal letter written by S/Ldr Haw to mother of P/O Pyle. *G.F. Pyle*

"Some of the Boys – Coolham 1944"

From left: F/O G.R. Dickson, F/Lt J.P. Bassett, F/O A.F. Osbourne, F/O S. Payne, F/O A.B. Thomson, F/Lt R.G. Kleimeyer, F/O M. Humphries holding Monty Bush. *Photo: L.G. Lunn*

Life under canvas for 129 Squadron, Coolham 1944.

P/O G.F. Pyle (left), F/Lt R.G. Kleimeyer (centre) and F/O D.C. Parker (right). *Photo: L.G. Lunn*

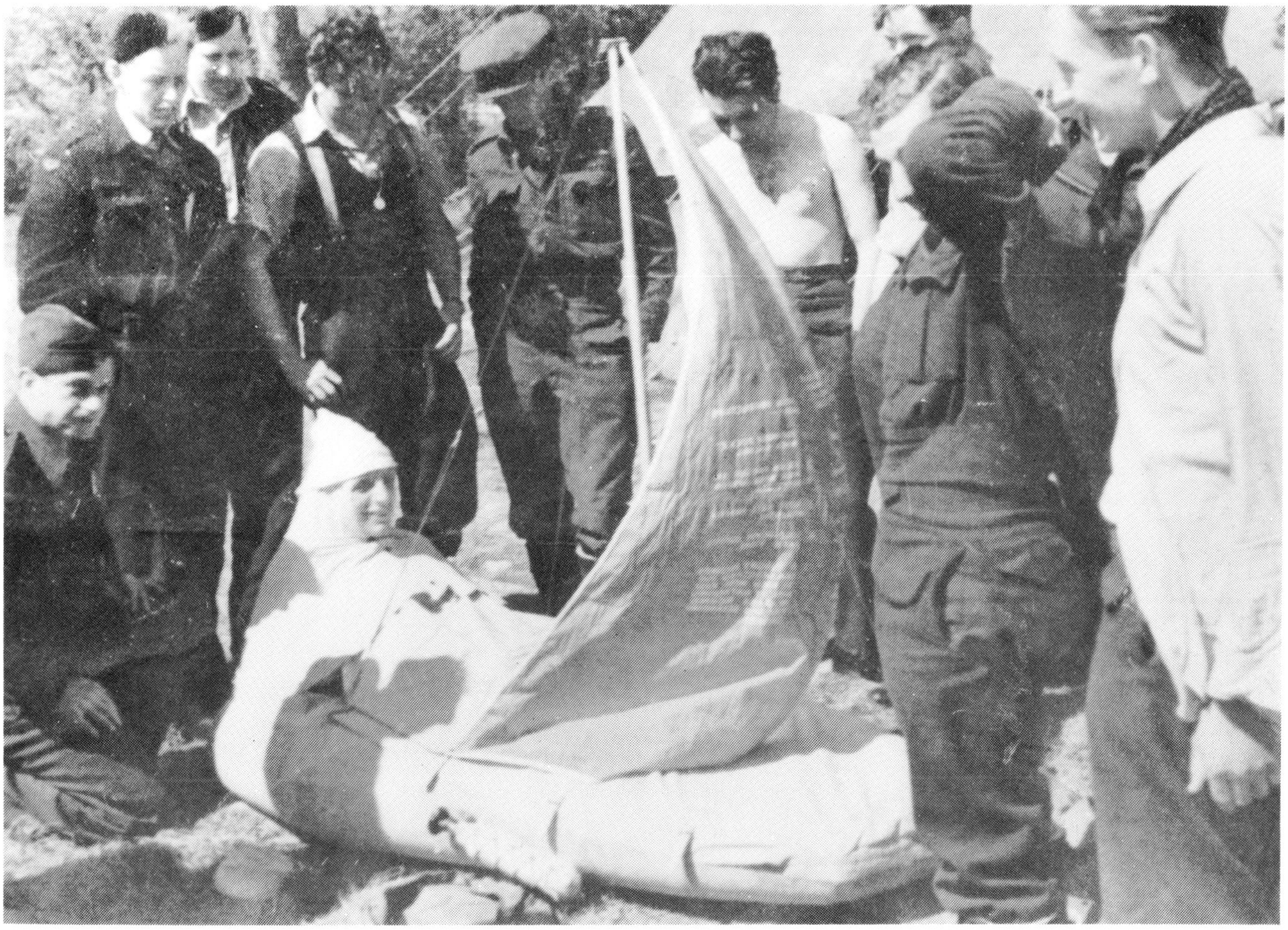

Groundcrew gather around F/O D.C. Parker of 129 (Mysore) Squadron prior to a dinghy race on Old House Lake, adjacent to the airfield. *Photo: J.P. Bassett*

P-51c Mustang at 129 Squadron dispersal. *Photo: J.P. Bassett*

Headstone to grave of F/O Stanley Payne, 129 (Mysore) Squadron. buried at Hills Cemetery, Horsham, West Sussex. *Photo: P. Hamlin*

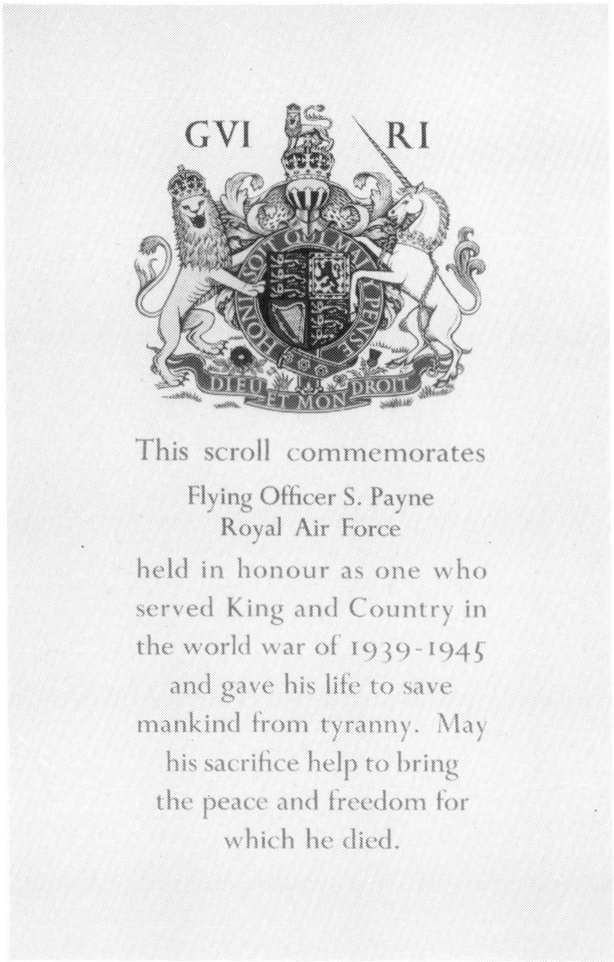
GVI RI

DIEU ET MON DROIT

This scroll commemorates

Flying Officer S. Payne
Royal Air Force

held in honour as one who served King and Country in the world war of 1939-1945 and gave his life to save mankind from tyranny. May his sacrifice help to bring the peace and freedom for which he died.

Commemorative scroll in memory of F/O Stanley Payne. *Reproduced by kind permission of Christine Shiner*

Jannie Bush, 129 (Mysore) Squadron mascot. *Photo: L.G. Lunn*

P-51c Mustang at 129 dispersal next to Old House Lake, Coolham. *Photo: L.G. Lunn*

The life and soul of 129 (Mysore) Squadron, F/Lt Richard (Guns) Conroy, RAAF.
Photo: R.J. Conroy

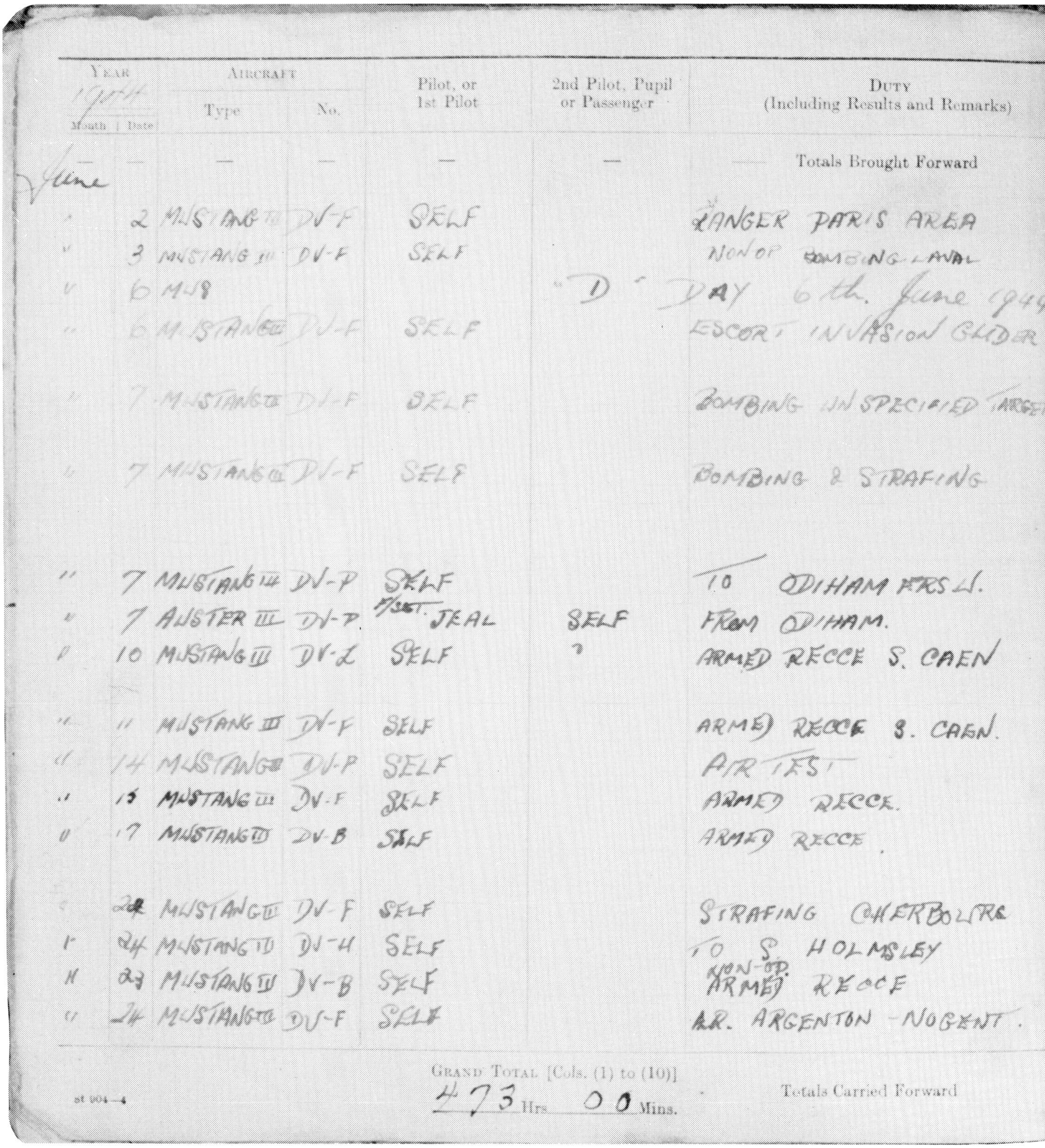

Year 1944 Month	Date	Aircraft Type	No.	Pilot, or 1st Pilot	2nd Pilot, Pupil or Passenger	Duty (Including Results and Remarks)
—	—	—	—	—	—	Totals Brought Forward
June						
"	2	MUSTANG III	DV-F	SELF		RANGER PARIS AREA
"	3	MUSTANG III	DV-F	SELF		NON OP BOMBING LAVAL
"	6	MUS		"D" DAY 6th June 1944		
"	6	MUSTANG III	DV-F	SELF		ESCORT INVASION GLIDER
"	7	MUSTANG III	DV-F	SELF		BOMBING UNSPECIFIED TARGET
"	7	MUSTANG III	DV-F	SELF		BOMBING & STRAFING
"	7	MUSTANG III	DV-P	SELF		TO ODIHAM
"	7	AUSTER III	DV-P	F/SGT. JEAL	SELF	FROM ODIHAM.
"	10	MUSTANG III	DV-L	SELF		ARMED RECCE S. CAEN
"	11	MUSTANG III	DV-F	SELF		ARMED RECCE S. CAEN.
"	14	MUSTANG III	DV-P	SELF		AIR TEST
"	15	MUSTANG III	DV-F	SELF		ARMED RECCE.
"	17	MUSTANG III	DV-B	SELF		ARMED RECCE.
"	22	MUSTANG III	DV-F	SELF		STRAFING CHERBOURG
"	24	MUSTANG III	DV-H	SELF		TO S. HOLMSLEY
"	23	MUSTANG III	DV-B	SELF		NON-OP. ARMED RECCE
"	24	MUSTANG III	DV-F	SELF		A.R. ARGENTON-NOGENT.

Grand Total [Cols. (1) to (10)] 473 Hrs. 00 Mins. — Totals Carried Forward

Flying logbook of F/Lt R.J. Conroy of 129 (Mysore) Squadron, detailing events of June 1944.

Single-Engine Aircraft				Multi-Engine Aircraft						Passenger	Instr/Cloud Flying [Incl. in cols. (1) to (10)]	
Day		Night		Day			Night					
Dual	Pilot	Dual	Pilot	Dual	1st Pilot	2nd Pilot	Dual	1st Pilot	2nd Pilot		Dual	Pilot
(1)	(2)	(3)	(4)	(5)	(6)	(7)	(8)	(9)	(10)	(11)	(12)	(13)
		6·05 1·45										
84·25	357·35											
	2·30	YELLOW 3 WEATHER CLAPPED OUT SO COULDN'T COMPLETE PLANNED TRIP. STOOGED AROUND SHOT UP 4 BARGES NR. NOYON NOTHING ELSE ABOUT.										
	·10	BOILED ON TAKE-OFF										
C		Invasion of France.										
	1·30	ESCORTED 72 DAKOTA TUGS & HORSA GLIDERS LANDING ON BEACHHEAD AT LE HAVRE LITTLE FLAK NO E/A OPPOSITION THOUSANDS OF ALLIED A/C AND SHIPS. INVASION GOING WELL										
	1·40	SWEEPING INVASION AREA FROM S. ROUEN TO ARGENTAN BOMBING ANY TARGET OF MILITARY IMPORTANCE. EVENTUALLY PRANGED WOOD AND SEARCHLIGHT BATTERY NO. E/A LITTLE FLAK.										
	1·50	FOUND A LARGE CONVOY OF M.T. INCLUDING AMMO FUEL & ARMOURED CARS ETC. FIRST ONE I CLOBBERED BLEW TRUCKS UP WITH TERRIFIC EXPLOSION A PIECE OF METAL PIERCED MY WING AND MADE A NICE HOLE. ALSO DAMAGED 4 OTHER M.T. TRUCKS. A WONDERFUL SHOW. NO FLAK NO E/A										
	·20											
									·20			
	2·15	RED 2. SWEEPING AREA TO FIND ENEMY TRANSPORT. EVENTUALLY BOMB SMALL COLUMN AND THEN STRAFED SAME GETTING DARK SO HAD TO LEAVE LANDED AT BASE IN DARK. ALL ROADS IN FRANCE LIT UP BY WRECKED AND BURNING M.T. LITTLE FLAK										
	2·15	RED 2. VERY LOW CLOUD OVER AREA VERY QUIET NO M.T. COLUMNS. BOMBED CROSSROADS VERY POOR SHOW.										
	·40											
	2·25	RED 2. ABSOLUTELY NOTHING ON ROADS, NO E/A, DEAD. DROPPED BOMBS ON RAILWAY. BLOODY POOR SHOW.										
	2·30	RED 2. TOOK OFF 5 AM. FOUND CONVOY OF TRUCKS NR. AINYE PRANGED WITH BOMBS AND STRAFED AT N. LE MANS FOUND ANOTHER CONVOY ON MOVE TOWING 88 MM. GUNS. STRAFED WITH EXCELLENT RESULTS 2 M.T. VEH. DESTROYED 3 DAMAGED.										
	1·45	RED 3. STRAFING AREA S. CHERBOURG. DIVED FROM 6000' BRITCHED UP ANY THING WHICH CAME IN SIGHT TERRIFIC LIGHT FLAK. JOE HIT IN STRAP AND										
	·30	LANDED ON STRIP. HANK SHOT DOWN IN FLAMES - DEAD. PRETTY SHAKY DO.										
	·40	NO ELECTRICITY										
	2·10	RED 3. BAGS OF HUNS OVER DRIEUX BUT DID NOT GET A CRACK AT THEM. NO FLAK. BOMBED RAILWAY - DIRECT HIT - NO E/A - DULL SHOW										
84·25	380·45	6·05 1·45										
(1)	(2)	(3)	(4)	(5)	(6)	(7)	(8)	(9)	(10)	(11)	(12)	(13)

R.J. Conroy

11

Squadron Leader Charlton "Wag" Haw, DFM, DFC, Order of Lenin, C/O of 129 (Mysore) Squadron

Squadron Leader Charlton "Wag" Haw, 117992, was Commanding Officer of 129 Squadron. He was an outstanding and natural fighter pilot having flown in the Battle of Britain. He was shot down in combat over Bristol in September 1940 after destroying a Bf 110. He landed unhurt.

In 1941, as a Flight Sergeant flying Hurricane fighters, Wag was sent to Arctic Russia with 81 Squadron forming part of 151 Wing. Their orders were to protect the ports of Murmansk and Archangel and to teach Soviet pilots and ground crew how to fly and maintain Hurricanes.

Wag brought down three Me109's making him the top scoring fighter pilot in the Wing becoming known as "The hero of Murmansk".

He was awarded the Order of Lenin, Russia's highest Order, one of only four given to the British in the war.

On his return Wag was commissioned, awarded the DFM and posted to a Spitfire squadron at Hornchurch. He was then given command of another Spitfire squadron at Biggin Hill before joining 129 which converted to the Mustang fighter bombers escorting Allied bombers on daylight raids and covering the D-Day landings. Altogether he survived four tours of operations.

Wag was highly respected by all the pilots at Coolham – not least for his boogie woogie piano playing – all the rage at the time. He kept his squadron entertained many an off duty hour.

"When he sat down at a piano it came alive and so did the piano," said F/Lt Joe Leigh.

Sadly, when Paul Hamlin traced Wag to his home in Farnham, Surrey, he was already terminally ill with cancer but Paul spent many a happy hour with him, reminiscing and tape recording his memories for inclusion in this book. His humour shines through.

Of his nickname, Wag said: "People have often asked me how I was christened Wag. It happened when a chap who trained with me who came from Yorkshire asked me my name and I said Charles or Charlie and he said everyone from Sheffield called Charlie is called Wag and from then my nickname stuck and everyone in the RAF knows me as Wag."

Wag died on 27 November 1993 aged 73, seven months before the anniversary weekend celebrations in Coolham the following June. But he and his second wife, Audrey, visited the site of the former airfield in September 1993 with Paul. They had a drink at the Selsey Arms pub where Wag saw the site of the memorial, fully approved its design and wished the project every success.

These are his recorded memories.

I was born in York on 8 May 1920 and always wanted to be a pilot so I joined the RAFVR at Hull before the war. I started flying early in 1939, finished initial training and had just started intermediate training when war broke out and I was called up immediately on 1 September 1939.

I finished advanced training in May 1940 and because of the fall of France went straight to 504 Squadron which had just come back from France.

They were then in Wick, Scotland and after familiarisation on Hurricanes we did various interceptions from there and then moved to Hendon for the Battle of Britain.

After that we went to Exeter at the end of September 1940 then to South Wales where the squadron was split in two. "A" Flight and the C/O were sent to form 81 Squadron to go to Russia. We left for Russia in August 1941 and flew for six weeks operational and then six weeks handing over to the Russians, coming back to this country on 7 December 1941, the day Pearl Harbour was attacked.

On the way back from Russia I heard on the radio I had been awarded the Order of Lenin which was presented to me by the Russian Ambassador to England at the Russian Embassy in London.

At the same time I was commissioned and also awarded the DFM.

I was then sent on rest supposedly for six months but

after six weeks I managed to get hold of my C/O who I'd flown with before the war and they – 81 Squadron – were moving south. I went back on ops and the wing moved to Hornchurch and after three months at Hornchurch as a Pilot Officer I was made a Flight Lieutenant in 222 Squadron which was just across the airfield. After four months as flight commander there I was made a Squadron Leader and sent to Biggin Hill to command 611 Squadron.

After short tours with them I left them in May 1943 to go on a six week factory tour, talking to the workers. I found that much more of a strain than actual fighting and I pleaded with the Air Ministry to get me off that before my health collapsed completely. I then managed to talk my way back into ops. and was given command of 129 Squadron, again at Hornchurch and we did various sweeps attacking shipping, all the things that fighter squadrons were doing at that time. Later on we were posted north to Peterhead where eventually we were to re-equip with Mustangs and early in 1944 we moved south to Heston and then eventually to this Advance Landing Ground at Coolham.

At the ALG we lived in tents and the Mess and dining room were large marquees which was a bit rough after the accommodation we had been used to.

When we became operational at Coolham we did several short missions to France, some with only four aircraft, having a go at anything that moved. We were also given bombs which we used to attack anything that we thought might be of strategic importance. There were very few actual targets given to us at that time and we had had no bombing practice whatsoever. We just did it by trial and error.

Then we did several escorts with the Americans to Germany. Those proved very ineffective as far as air fighting went because by that time the Germans were doing head on attacks and it was impossible to catch them. They came in at a closing speed of perhaps 700 miles an hour and there was a flash of an aeroplane and it was gone. It was just one attack and then it dived away. There was a lot of flak and that was the most uncomfortable thing, avoiding flak over these heavily defended targets.

It was either daily escort with the Americans doing this long range bombing or these attacks in France in small groups attacking various selected targets and, occasionally, with bombers to France.

The longest trip I did was to Berlin which took about five hours. Coolham was fairly easy to find if the weather wasn't too good because coming across the downs if you could find Chanctonbury Ring and headed due north you eventually came to Coolham and that's the way many times I found my way back to the airfield.

D-Day, 6 June, from a Wing point of view, was disappointing. We sat around most of the day as we weren't called out to fly until 8 o'clock in the evening which was as escort to some bombers that were bombing targets inside France round the eastern sector of the invasion area.

One of my pilots shot down an Fw190. That seemed to be the only enemy aircraft around. There were plenty of friendly ones.

After D-Day we still carried on with these small, short intrusions into France and escorting the Americans.

A typical day would be to get down to dispersal at dawn unless you knew the day before that the squadron wasn't needed until later in the day. Then after you had been given your orders or flight or wherever you were going to go you'd be debriefed by the squadron intelligence officer who would make his report and which would be sent off to Wing and later to Fighter Command or Group Headquarters in this case. A lot of the time was spent hanging about reading, playing darts and really at times it could be quite boring.

Regarding my ground crew, they were all Polish and they had a distinct hatred of the Russians as well as the Germans and quite frankly if you came back sometimes without having fired your guns they were more than disappointed, thinking perhaps you hadn't been looking hard enough.

I don't know who thought of putting me with a Polish Wing, being the holder of the Order of Lenin, but whoever it was must have done it with tongue in cheek, I think, for I remember an incident when one night, coming back from the pub, I was walking along and a very large Polish Flight Lieutenant got hold of me and said: "Come here my little Russian Sir," and quite frightened me. But he was eventually calmed down by some fellow officers and I believe – I didn't see him again – he was posted away the following day when the incident had been reported, not by me but by their own people. We got on well with the Poles generally. They were a very tenacious bunch.

I was 23 at the time and the C/O of one of the Polish squadrons, Horbaczewski, was 27 which I possibly thought was an old man in those days because a lot of us who had flown in the Battle of Britain were only just 20, so four years' difference seemed like a long time. Although we didn't have a lot to do with each other we had the occasional drink together.

In the air we all got on extremely well, although at times, if there was a mix up or a bit of a scrap, a dogfight, the Poles would quite naturally break into their own tongues when they got very excited. Then the English pilots didn't know what the hell was going on and there would be screams of not very nice language from the English saying "speak f...... English you bastards" – which usually produced some sort of a result.

Having about 28 pilots in the squadron and only 12 taking part at a time we had to rotate their names and this was the duty of the flight commanders, to give everybody a chance to have a go.

Usually, if the Wing leader was leading the squadron and the squadron commander didn't fly then, to give them experience, occasionally Flight Lieutenants would be given the chance to lead the squadron on certain operations.

Of course, we were completely controlled by the weather. Each squadron would be given the time to be at "readiness" which meant the aircraft were to be ready to fly immediately, or some squadrons would be on standby which meant half an hour and some on an hour's readiness.

If the weather was good then orders would come through from Fighter Command or 84 Group giving us various tasks to perform for the day and, of course, some days we didn't fly at all, and those days it was rather difficult to know what to do with people.

Sometimes I believe we went on route marches, had football matches, marched to the nearest pub and back, that sort of thing. For our personal transport we were given a jeep, a small four seater, four wheeled drive machine and it's really surprising how many people you can get in a jeep when you want to get down to the pub in the evenings. Usually you could get 10 or 11 hanging on in. I don't think it would go down very well today but it certainly got us there and back.

There wasn't very much to do locally apart from play darts, drink a few beers then come back, settle down for the evening, then up early, depending on whether or not there was a sweep or attack planned for that morning.

There was one rather amusing incident one day when I wasn't flying and the Wing was taking off. The latrines, which were a lot of wooden poles wrapped round with hessian, were well to the side of the runways and on take off one of the Mustangs burst a tyre and slewed towards the lavatories and the wing of the aircraft knocked the lavatory down leaving this Polish person sitting on this Elsan. I don't think he was constipated after that, the noise must have been tremendous. I can see him sitting there now.

One of the things that happened towards the end of 1944 was when we were on a long distance flight and one of the pilots in some Wing must have left his transmitter on and as we were over Holland he was hit and all the way down he was screaming for his mother, and that was most unnerving, and then silence as he went in. But he must have just left it on or perhaps if he was hit, presssed it, but all the way down, very unnerving.

And we were in cloud all the time. It was a useless thing, he must have been hit by flak, and you couldn't see anything. A lot of these things were an absolute waste of time. You knew when you went over there you had no hope of seeing the target.

It is difficult after a long time to remember the five who didn't make it from Coolham in my squadron but one was F/Lt Green, whose parents lived in Worthing and were in the catering business in a large way.

We used to go to his house for a bath now and again and when he was missing I went to tell Mrs Green personally. She opened the door all smiles and said: "Come in Wag. Have you come for a bath?"

And I had the most terrible business of explaining to her what had happened. We saw him crash but then I don't know what happened after that.

F/O Hancock was another. I had gone to Skip Payne's funeral. Skip was killed when his engine failed and he crashed in some trees at Coolham.

Hank was leading our squadron, 129, and was hit when they crossed into France while I was at the funeral and I, in a way, feel a little bit responsible for his death because although I might not have crossed the coast at the same place I always think of Hancock as probably dying for me, in a way.

There was also Nelson, who I think was probably shot down on one of these low level attacks. A lot of these good pilots were shot down by these "rhubarbs" or whatever they were called in those days, by light flak.

Really, to my mind, the amount of damage done I didn't think it warranted the risk in some of these cases. Of course, when anyone was missing the C/O usually wrote a letter to the parents or relatives of the missing person and it was a very upsetting business to have to write these letters.

In the case of some of them, of course, they did come back, George Pyle for instance. I had written a letter about him which George told me the other day he still has and fond memories of what I thought about him.

Another incident was when four of us had gone over to France to dive-bomb some crossroads or any targets we could find and when I dived down only one bomb came off. I immediately told the others to follow me back to England and over the Channel I tried to get rid of this bomb by various means, rolling, diving, pressing all the switches, but nothing would happen so I called up Tangmere.

They said: "Stand by and don't land until we've cleared the airfield."

So I stood off Littlehampton, still trying to get rid of it and eventually Tangmere said: "It's Okay now, you can come in."

I had two or three more goes at getting rid of it and eventually it did come off and fell just off the beach at Littlehampton without exploding.

Although the Spitfire and the Mustang were powered by exactly the same engine they were quite different aircraft. The Spitfire was very nose-heavy, very narrow undercarriage but in the air I would say nicer to fly than the Mustang. It was much more manoueverable and had a much better turning circle than a Mustang.

A Mustang, however, particularly for flying from rough airfields, had a much stronger undercarriage and was a faster aircraft. Particularly in a dive, in some cases they experienced what they call compressibility whereby they were approaching speeds when one had to reverse the controls in order to regain control of the aircraft again. That happened several times but I don't think anyone was killed because of that.

The Mustang had a better rolling capacity than the Spitfire but of the two aircraft I enjoyed flying Spitfires more than the Mustang.

Towards the middle of June my tour of duty was coming to an end. We were never told that we were going to be taken off and sent on rest, you were called before the Wing Commander, Flying, or if you were more junior rank, the squadron commander, and told that you were grounded from operations on that day and sent to some other flying appointment.

My turn came on 22 June when the squadron was posted to Holmsley South and on that day we were taken off offensive operations and put on to chasing VI's, the flying bombs, which the Mustang could catch if it had a slight advantage of height, so I lost the opportunity of having a go at them.

From that day I was posted to the Central Fighter

Establishment where we trained pilots for operational duties. That was at Millfield in Northumberland.

After I finished my stint of duty at Central Fighter Establishment, which by then had moved down to Tangmere, I was offered a Wing of Mustangs which were flying from Scotland on long range escort to Mosquitos but the war then was nearing its end and I never took up that appointment.

After the war, in 1946, I was given command of a fighter squadron again, 65 Squadron, flying Hornets, until 1948.

I then had over two years on the air staff of 64 Group at York but then I was caught out on my annual medical and failed the eyesight test. They gave me the option of resigning early or transferring to one of the ground branches. Having flown for so many years I decided to leave the RAF, retiring in 1951.

Eventually I finished up quite near Coolham at a pub called the Blacksmith's Arms at Adversane. We stayed for six years from 1952 to 1958.

The Blacksmith's Arms was a very old pub which had been run down rather badly but I had a lot of fun hacking it about exposing the beams, old fireplaces and putting it back to what it should have been. There was a very good passing trade but not a lot of local trade as there were only about a dozen houses in the village of Adversane.

I put it down to a six year working holiday. We came out of there with the same amount of money in the bank as we did when we went into it but we had a hell of a six years. I shall never forget it.

After that I went in the petfood business where I stayed for 26 years until I retired when I was 65.

Looking back on it all I suppose one could really say that if you wanted, as I wanted, to be a pilot, particularly a fighter pilot, it was just a schoolboy's dream.

While one had periods of boredom flying you also had moments of sheer terror which blended in quite nicely.

When Paul got in touch with me about this commemorative day at Coolham it was such a long time ago it is very difficult to remember all the things that happened there, but one can say one had an enjoyable time there, I suppose, apart from the poor chaps that didn't make it.

I do wish Paul and his committee well on such a worthwhile project as this because there must be many other airfields around that haven't had the time and trouble taken over them to put them back on the map again.

This will in some way help to remember the people that didn't make it from that Advance Landing Ground and I wish you good luck with it Paul.

I was decorated with the DFC, DFM and the Battle of Britain Clasp, the Air Efficiency Award, the Order of Lenin and the Russian Victory Medal which was given me in Moscow in 1985 when I and my wife were invited there to attend the celebrations marking the 40th anniversary of the end of the war, plus the Atlantic Star and several other campaign medals. I destroyed seven aircraft and several other probables and, of course, damage to shipping and armoured vehicles, really too many to count.

F/Lt Joe Leigh of 129 said: "Wag led well. He was a good squadron commander and he was a good Squadron Leader and there is a difference. He led us well on all our shows."

Squadron Leader Charlton "Wag" Haw, DFM, DFC, Order of Lenin. *Photo: L.G. Lunn*

Coolham, 29 May 1944. S/Ldr Haw speaking with General Sosnkowski. Also present, S/Ldr Horbaczewski (left) and S/Ldr Lapka. *Photo: J. Bargielowski*

Charlton "Wag" Haw playing boogie woogie at his home in Surrey a month before his death on 27 November 1993. *Photo: J. Sampson*

12

222 Natal, RAF Squadron

This squadron's motto was "Pambili bo" (Zulu), "goes straight ahead". The crest shows a wildebeest in full course. It is taken from the armorial bearings of Natal, 222 being a gift squadron.

Flight Lieutenant Ross Oliver relates:

It usually contained a couple of pilots from different Commonwealth countries as well as favoured foreign Allies – Canada, New Zealand, Australia, Rhodesia, South America, Northern Ireland and Norway were all represented when we were at Coolham.

I joined 222 at Hornchurch in September 1943 having trained on Tigers and Wirraways in Australia and arriving in England early 1942.

My Above Average rating didn't appear to count and I was sent off to do a full course, finally getting to 61 OTU Rednal in July 1943.

I was fortunate to meet another "colonial" at Rednal, "Screwball" Beaurling, who was my instructor. We spent many off duty hours skeet shooting and I learnt a lot from him.

He was a hero and the greatest character one could ever know. With his help I easily topped the air to ground and air to air shooting with Spit Mk I and Mk II's and again a rating of Above Average.

Screwball got away with murder, so to speak. Wandered about with his trousers rolled up half way to the knees, carried a shotgun everywhere he went, and by the barrel, just to be different.

One day, after morning break, when the girls had retired from the NAAFI van and were nicely settled in the female toilets, he let both barrels off under the eaves of that building. I have never seen so many half naked females exit from a building, and in such haste, in all my life.

But nobody would take action against him, much too popular. A real character with whom I spent many a happy hour.

With 222 Squadron I moved to various airfields and eventually to Coolham.

My records of our brief stay here are almost identical to other information sent to me. I have noted the bombers came under intense anti-aircraft fire. Those sorties were purely routine, as far as my memory allows.

I recorded 500 hours of service flying a few days after Coolham with over 400 hours on Spitfires.

John Lenehan, also RAAF, and myself, were commissioned at the same time. My Officer's Certificate of Service shows the appointment as 27 June 1944.

Of all the people in my life I hold the memory of Ernie Broad with the greatest respect. He was just back from Malta (F/Lt, DFC) and my O/C Flight, when I joined 222 Squadron.

Recently both married, a natural bond developed at once which lasted until his death at the end of 1991. He finished as a Squadron Leader with a Bar to his DFC and a credit to himself and his family in Easingwold, north of York, where he was Mayor for many terms and gained the respected name of "Mr Easingwold".

On looking back, considering 222 Squadron was made up of pilots from different Commonwealth and other countries, it is strange how it was only the English pilots who were considered for promotion, who were always appointed as C/O of the squadron and O/C of Flight.

It is also strange how it was only the English who gained an award and yet we all did the same job. I know from what Ernie Broad told me before he died that I and a couple of others were recommended for awards, but for some reason he couldn't understand why they failed to be gazetted.

It is not "sour grapes" that I now mention these matters but it does nothing for the feelings that once existed to know, so late in life, to be so close and yet not be recognised.

I am, however, quite philosophical about these matters. I know what I did, many dangerous sorties from which I may not have returned, but luck was mostly with me.

Most of what I say is said with "tongue in cheek" but at the same time some truth remains.

It is also said that "stories" from old ex-soldiers, and in this case ex Air Force types, remain just that, a "story" and little of what is told should retain any belief.

For that reason I have been reluctant to talk of my experiences, as some are almost unbelievable. I just can't imagine I was once so "hare brained" to do such things.

Please do not be offended, but the ties that once bound our two countries so strongly have almost broken.

Our present Prime Minister appears determined for our country to break all ties and become a Republic.

Many other things have caused the feelings to change. The atom bomb testing in Australia and the refusal by Britain to accept responsibility and help clean up the mess.

The Common Market, leaving us with little time to find alternative markets, hasn't helped, also the recent Royal happenings.

But the greatest insult was the Home Office announcement that Australia were certainly in the "Second Eleven" when it came to the Normandy 50th year celebrations.

The Home Office considered none of the "Colonial" countries should be involved, only Britain, USA and France. That didn't go down very well at all.

They forget that over 22,000 Australian airmen were involved in actual operations over Europe from Britain and that more than 6,000 died.

They completely forgot to mention how many French Divisions were involved in the Normandy landings or what percentage of the French got into bed with the Nazis during the occupation.

I will be forever thankful I was able to attend a reunion with five other members of 222 Squadron and our wives in 1991 in Bournemouth, England.

Bill Gower (Canada); Charles Ernest (Ernie) Broad (S/Ldr, DFC and Bar), from Easingwold, near York; Bob Beardsley from Woodbridge in Suffolk; Robert H. Reid from Chesterfield; Harry E. "Tony" Turney, who was very ill at the time, from Ferndown, near Bournemouth and myself from Australia.

Both Turney and Beardsley finished with a DFC and the rank of Squadron Leader.

It was a most wonderful reunion after all those years. Sadly, within a year, Bill Gower and Ernie Broad had died within weeks of each other and shortly afterwards Tony Turney also died after a long illness.

This hit me very hard. These men were almost immortal to me and I had not prepared myself for the loss of such close and good friends.

In 1994 the RAAF put together a video, *The Ross Oliver Story*, of my time in the Air Force, which has now gone into the archives.

Yes! It is rather nice to be recognised after all these years and to know it was appreciated.

Ross Oliver, W/O, AUS 408314, lives in Victoria, Australia, with his wife, Sylvia, a WAAF he met and married in Nottingham in 1943.

❑❑❑❑

Squadron Leader Bob Beardsley, a Battle of Britain veteran and a pilot in 222 (Natal) Squadron, recalls a "sortie" on 1 July 1944 which he led.

We escorted the bombers out over target and back again against attack by Luftwaffe fighters. Although the weather was poor on take off we made the rendezvous and carried out our mission – very disappointing!

The bombers bombed through 10/10th cloud cover, if I remember rightly. We were busy hoping for targets but no hostile fighters appeared so we brought the bombers home – a very unsatisfactory sortie, I'm afraid and nothing exciting to write about or remember – except we were glad to find good weather over Coolham on our return!

Sorry I can't stretch this to something more interesting but a lot of fighter escorts were "boring" and we were too alert for enemy attacks then to look at the ground for long!

I only spent six months in the squadron before being pulled out and posted to a staff job in Cairo, much to my disgust!

Robert Arthur Beardsley, DFC, F/Lt 100607. He lives in Suffolk.

❑❑❑❑

W/Cdr David Cox, a Battle of Britain veteran, commanded 222 (Natal) Squadron as Squadron Leader. He arrived at Coolham ALG on 30 June after a 20 minute flight from Selsey Airfield.

There were no operations so I led 222 Squadron into Horsham and to the Royal Oak in Rusper, where the local beer was found to be good. This fact is recorded in my log book.

On 1 July the squadron did a fighter swoop near Paris. I did not go on this op. as I had been warned by Group HQ that I had done too many operational hours in June.

I took the opportunity to visit the American Air Force unit at Duxford where I had trained on Thunderbolts the previous August. The weather on 2 July prevented any ops. It also delayed my return from Duxford. In fact, I had to divert to Gatwick and wait for the weather to clear at Coolham.

It was a nostalgic visit to Gatwick as the last time I had been there was in August 1939 flying Tiger Moths in my training days. Now I was landing there in a Spitfire Mk IX as C/O of a famous fighter squadron!

July 3 and another visit to Horsham! July 4 saw an early departure to our new airfield at Funtington, near Chichester.

Despite being short of our groundcrews we did two ops, one of 1hr 30mins and the other of 2hrs.

My main memory of Coolham is of the lovely Sussex countryside.

David G.S.R. Cox, DFC and Bar, C de G, S/Ldr 101041 Reading.

❑❑❑❑

Leading Aircraftman Ron Clark was a fitter with 222.

I came to Coolham airfield in 1944. While there I carried on my normal duties as fitter 11A on Spitfires.

We were only there a few days, on one of which some of us returned to Selsey to clear up outstanding work there.

Although I lived at Wisborough Green, some seven or eight miles away, I was unable to get home.

We moved from Coolham to Funtington.

Ron Clark, LAC 1426444
Billingshurst, West Sussex.

❑❑❑❑

Awful weather colours the memories of Coolham of Flight Sergeant Pete King, an armourer.

As regards my time at Coolham the memories have faded. What I do remember is it was a lousy June in every respect. The weather was bloody awful – fog, lots and lots of mud, rain – a most depressing situation altogether.

I remember I drove a Thornycroft truck loaded with bombs and ammunition and thought to myself, I must be some sort of prat. I could easily have detailed someone else to drive.

I also remember you needed a parachute to descend from the cab, they seemed so high off the ground.

My co-driver was a Cpl Ginger Galley. Funny thing was, he could not drive but I took him because he seemed a real hard lad, just the chap to have if things got rough.

I recall one little incident when Ginger and I debated if it was best to sleep in the cab or under the truck. With hindsight, that begs belief, doesn't it, but as armourers we were trying to work out whether the blast would be stronger going up or down. Luckily, it was never put to the test.

I wouldn't have missed that part of my life and, in common with most of the men there, we saw fun and fear there together.

At the end of the day I look at the state the country is in today and wonder, was it all worth it!

Pete King, F/Sgt 552366
Cambridgeshire.

An apt slogan on the bomb amuses F/Sgt Pete King, an armourer with 222 (Natal) Squadron.

Photo: P. King

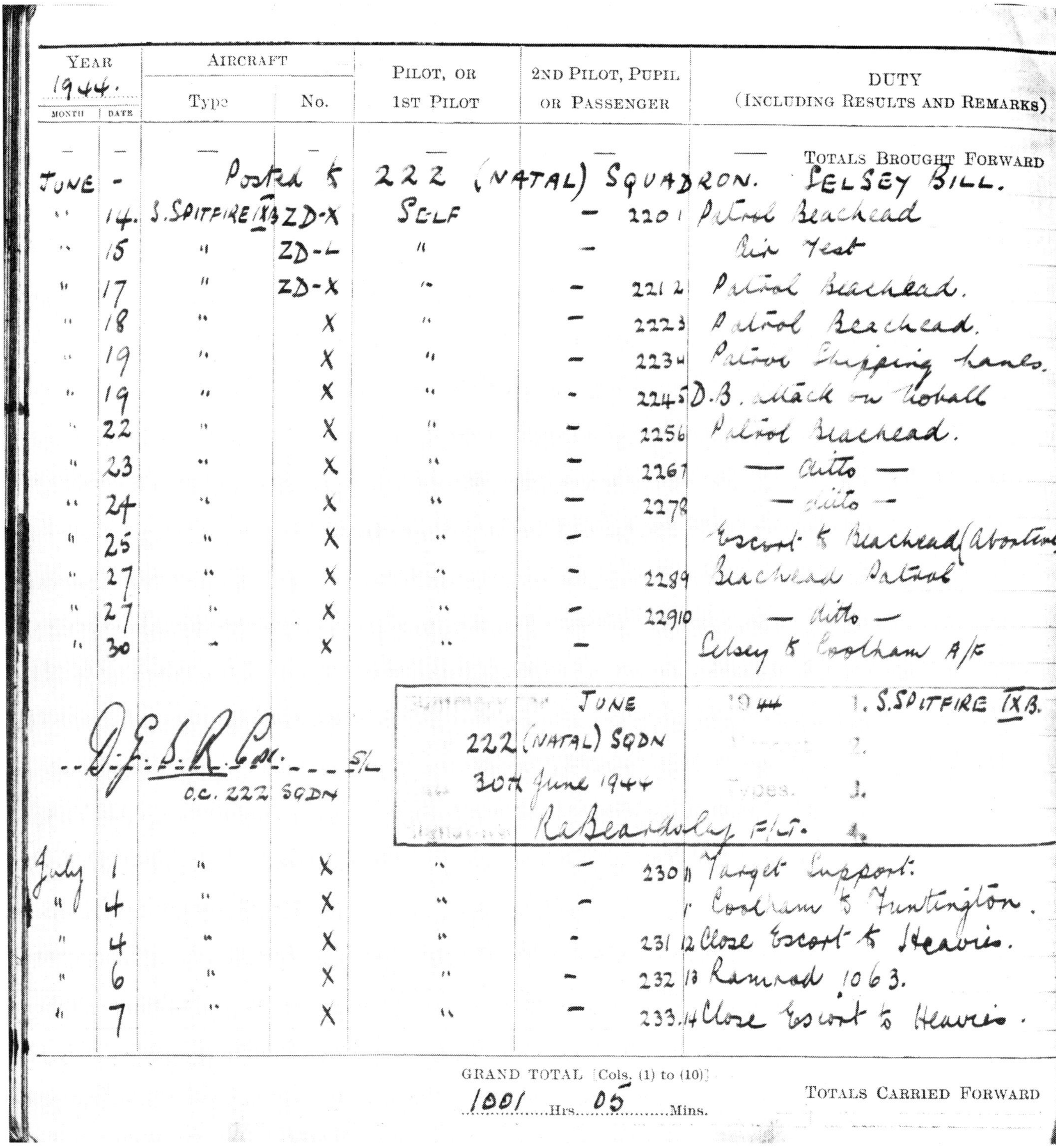

YEAR 1944. MONTH	DATE	AIRCRAFT Type	AIRCRAFT No.	PILOT, OR 1ST PILOT	2ND PILOT, PUPIL OR PASSENGER	DUTY (INCLUDING RESULTS AND REMARKS)
—	—	—	—	—	—	TOTALS BROUGHT FORWARD
JUNE	-	Posted to 222 (NATAL) SQUADRON. SELSEY BILL.				
"	14.	S. SPITFIRE IXB	ZD-X	SELF	—	220 1 Patrol Beachead
"	15	"	ZD-L	"	—	Air Test
"	17	"	ZD-X	"	—	221 2 Patrol Beachead.
"	18	"	X	"	—	222 3 Patrol Beachead.
"	19	"	X	"	—	223 4 Patrol Shipping Lanes.
"	19	"	X	"	—	224 5 D.B. attack on Uphall
"	22	"	X	"	—	225 6 Patrol Beachead.
"	23	"	X	"	—	226 7 — ditto —
"	24	"	X	"	—	227 8 — ditto —
"	25	"	X	"	—	Escort to Beachead (Aborting
"	27	"	X	"	—	228 9 Beachead Patrol
"	27	"	X	"	—	229 10 — ditto —
"	30	"	X	"	—	Selsey to Coolham A/F

D. J. B. R. Col... S/L O.C. 222 SQDN

Summary for JUNE 1944 — 222 (NATAL) SQDN — 30th June 1944 — R A Beardsley F/Lt. — Types: 1. S. SPITFIRE IXB. 2. 3. 4.

MONTH	DATE	Type	No.	PILOT	2ND PILOT	DUTY
July	1	"	X	"	—	230 11 Target Support.
"	4	"	X	"	—	Coolham to Funtington.
"	4	"	X	"	—	231 12 Close Escort to Heavies.
"	6	"	X	"	—	232 13 Ramrod 1063.
"	7	"	X	"	—	233 14 Close Escort to Heavies.

GRAND TOTAL [Cols. (1) to (10)] 1001 Hrs. 05 Mins. TOTALS CARRIED FORWARD

Flying logbook of F/Lt R.A. Beardsley, DFC.

135 WING AIRFIELD SELSEY BILL SUSSEX.

SINGLE-ENGINE AIRCRAFT				MULTI-ENGINE AIRCRAFT						PASS-ENGER	INSTR/CLOUD FLYING [Incl. in cols. (1) to (10)]	
DAY		NIGHT		DAY			NIGHT					
DUAL	PILOT	DUAL	PILOT	DUAL	1ST PILOT	2ND PILOT	DUAL	1ST PILOT	2ND PILOT		DUAL	PILOT
(1)	(2)	(3)	(4)	(5)	(6)	(7)	(8)	(9)	(10)	(11)	(12)	(13)
86.10	848.15	3.55	19.45	–	–	14.00	–	–	–	4.25	12.45	4.50
"B"	FLIGHT. COMMANDER –											
	1.40		Flew No2 to Co. heavy acc. light flak – no damage.									
	·15											
	2.30											
	2.15											
	1.40		Mid Channel to Normandy.									
	1.25		Pas de Calais NE Le Touquet.									
	2.00		Uneventful									
	2.25		– do –									
	2.10		– do –									
	·35											
	2.15		– do –									
	1.50		– do –									
	·20											
	37.50.		Ops. 20.45. TOTAL OPS. 281.20.									
	–											
	–											
	–											
	2.00		Lancs & Halifax's bombing noballs in Pas de Calais									
	2.20		Patrol from Rouen - Amiens - Abbeville. Heavy flak.									
	1.30.		Lancs & Halifax's on noballs in Dieppe area, no joy.									
	2.15.		Close Escort to 30 Mitchells on P.O.L. installation at Chartres.									
	1.35.		APPROX 700 L & H's. bombed hell out of CAEN at 7,000' Colossal !! Heavy flak, heavy artillery barrage very interesting.									
86.10	877.15	3.55	19.45	–	–	14.00	–	–	–	4.25	12.45	4.50.
(1)	(2)	(3)	(4)	(5)	(6)	(7)	(8)	(9)	(10)	(11)	(12)	(13)

R.A. Beardsley

222 (NATAL) RAF SQUADRON

Fitters and Riggers 1944

Back row, from left: LAC N. Shaw, LAC R. Dartnall, LAC A. Higton, LAC W. Holiday, LAC E. Parsons, LAC J. Tait, LAC T. Wright, LAC R. Barnes, LAC J. Toft, LAC A. Allen, LAC K. Scott.
Seated: LAC E. Goddard, LAC W. Howard, Sgt E. Chadwick, F/Lt W. Taylor, F/Sgt J. Hess, Cpl E. Smith, LAC A. Pinfold, Cpl W. Lee.
In front: LAC R. Clark, LAC T. Endean, LAC F. Oxlade. *Photo: R Clark*

13

349 (Belgian) Squadron

Squadron Leader I.C. du Monceau de Bergendal, DFC and Bar, C de G, held the distinction of commanding 349 (Belgian) Squadron, the second Belgian unit in the RAF, from inception in January 1943 until July 1944.

F/Sgt Freddie Moureau and F/O Pierre Erkes only recollect atrocious inclement weather and muddy operating conditions during their very brief and active five days at Coolham. It would be true to say that personnel were pleased to vacate the airfield on 4 July.

In October 1946, 349 Squadron transferred from the RAF to the Royal Belgian Air Force. Today it operates F-16 fighters from Beauvechain, Belgium, Lieutenant Colonel P. Wouters being Officer Commanding.

349 (Belgian)Squadron. *Photo: Musée Royal De L'Armée Et Histoire Militaire, Brussels*

Place	Date	Time	Summary of Events	References to Appendices
			sections landed in France for refuelling.	
Selsey.	28.6.44		Squadron was awakened at 06.00 for an operation to take place at 07.30. This was cancelled before 07.00 and pilots after having breakfast returned to bed. A grim day with rain, low clouds and drizzle. A short release gave the opportunity of visiting a cinema in Chichester. The news that we are going to a new airfield near Horsham is confirmed. In the evening another operation was cancelled due to bad weather.	
	29.6.44		A day with varying weather sees '349' twice over Normandy, once under bad weather conditions and once in brilliant sunshine. No enemy aircraft were seen but some flak was experienced. F/O Balasse, late of 349 Squadron, paid us a visit.	
	30.6.44		The month of June ended by 135 Wing moving to Coolham. In spite of the move 349 Squadron was fully operational and were twice over Normandy. On the first sortie the C.O. was very nearly shot down by the flak experts of Caen. Second sortie was uneventful. The day ended at Coolham where pilots erected their tents and made themselves as comfortable as possible. It soon started to rain.	
			Squadron Leader, Officer Commanding, No.349 (Belgian) Squadron.	

R.A.F. Form 540

See instructions for use of this form in K.R. and A.C.I., para. 2349. and War Manual Pt. II., chapter XX., and notes in R.A.F. Pocket Book.

OPERATIONS RECORD BOOK

Page No. 1

of (Unit or Formation) No.349 (Belgian) Squadron. No. of pages used for day

SECRET.

Place	Date	Time	Summary of Events	References to Appendices
Coolham.			349 Squadron started the month of July at Coolham and moved to Funtingdon on the 4th.	
	1.7.44		'349' settles down at Coolham in very bad weather. The Squadron escorted 300 Halifaxes over France where no opposition was met. We hear the welcome news that F/Sgt H.Limet, missing on 28th April, 1944, is safe but prisoner of war.	
	2.7.44		Bad weather in the morning clears after tea. Squadron flies to Normandy where due to weather it spends the night. Transport was obtained for Bayeux where some of the boys tried to get a few drinks. Welcome was suspicious due to the late hour and to the fact that the natives mistook R.A.F. Blue for Luftwaffe Blue.	
	3.7.44		Again bad weather. Squadron returns around 17.00 hours after a couple of sorties from Normandy. Weather was so bad there that sorties had to be interrupted. After dinner the Squadron took off again to escort Mitchells bombing ARGENTAN beyond the bridgehead. We shall be moving again tomorrow to Funtington, near Chichester, but this move did not stop operations.	
Funtington.	4.7.44		On American Independence Day the weather improved greatly. Squadron moved by 11.00. The afternoon required no less than 26 sorties from the Squadron in addition to pitching tents, collecting water, unpacking, feeding etc. Funtington seems a much more desirable place than Coolham. It has a concrete perimeter track and excellent runways. Our "Doc", F/O DeGand, tonight experienced the joys of a drunken stupor.	

Extract from 349 (Belgian) Squadron, Operations Record Book, RAF Form 540

Public Record Office Ref: Air 27/1744

14

485 (New Zealand) Squadron

This squadron was only briefly at Coolham but Flying Officer Max Collett has these personal reminiscences of Squadron Leader J B (Johnny) Niven, DFC.

I had the pleasure and honour of serving under S/Ldr J.B. Niven for the complete term of his leadership of 485 (NZ) Squadron, the New Zealand Spitfire Squadron. It was my first operational posting after completing my training in New Zealand, Canada and England.

Johnny Niven joined the squadron as its Commanding Officer at Hornchurch in February 1944. He replaced a very popular officer, S/Ldr M.R.D. (Marty) Hume, a no-nonsense typical New Zealander and a part Maori.

Johnny's immediate problem, as a Scot, was to overcome the handicap of being the only non New Zealand pilot to join the squadron but such was his personality, flying ability and personal manner that he soon gained the pilots' respect and loyalty and led the squadron through more than 1,100 operational sorties without losing a pilot due to enemy action.

During his term as Commanding Officer, the squadron operated from Hornchurch, Selsey, Coolham and Funtington in England and Carpiquet, Eu (Le Treport) and Merville on the continent.

He led the squadron through the exciting days of D-Day and the subsequent Normandy battles during which the squadron had the honour of shooting down the first German aircraft over the beaches on D-Day.

He was replaced by S/Ldr Johnny Pattison, a Battle of Britain pilot from my own home town in New Zealand.

I personally did not fly from Coolham airfield as a week before I had flown from Selsey to Newchurch to meet a friend from New Zealand and broke an ankle when jumping down from my Spitfire after landing.

I experienced the ultimate disappointment of listening to reports of the invasion from a hospital bed at Halton Hospital. Fortunately I rejoined the squadron in time to participate in the historic move to a French airfield – Carpiquet near Caen in Normandy.

Johnny was an accomplished pianist and needed little encouragement to lead the men, accompanied by W/Cdr Herries, in many a boisterous and often bawdy sing-song. Maori songs caused him a problem sometimes but he soon picked up the rhythm. At times he had a little trouble coping with the New Zealanders' individuality and their total disregard for Air Force "Bumph" and unnecessary authority.

He was often heard to ask the station commander, "What did they break last night" or "How many are sober enough to fly today"? However, he survived and came through with flying colours.

I am sincerely grateful for the help and encouragement Johnny gave to me personally and it was due to his recommendation that I received my commission while at Merville and before he left the squadron.

I am pleased to be able to pay my small tribute to a very fine officer, who died several years ago, and to congratulate Paul Hamlin for the wonderful work he has done in perpetuating the memory of Coolham airfield and all who flew from it.

M A Collett, W/O NZ-422260
Secretary of 485 (NZ) Squadron Association
Napier, New Zealand.

One mission when his No 2 "vainly pursued" a fast disappearing Fw190 is recalled by Squadron Leader K.C. Lee.

During the Normandy invasion I was flight commander to "B" Flight of 485 and according to my log book I first landed at Coolham on 30 June after a beachhead patrol flying from the beachhead airstrip B2.

According to my rather vague memory we had spent the night at this Normandy airstrip and had been scrambled in pairs to carry out a very early patrol before dawn over the American beaches Omaha and Utah.

I remember that as we passed over the American beaches at about 3,000 feet, while still dark, it seemed that every 20mm and 40mm gun in the beachhead opened up with considerable malice directed at me and my No 2.

In hindsight I believe that there were Hun aircraft in the

area at the time and that we were not recognised as friendly. Nevertheless, at that moment we did not feel charitably disposed towards our American allies and in the dark and ensuing evasive manoeuvres my No. 2 and I became separated.

When, however, I called my No. 2 on the radio he assured me that he still had me in sight but was having difficulty keeping up with me.

Since it was now beginning to get light for my part I was unable to see any sign of him and I was certainly not going anywhere at high speed. Consequently I told my No. 2 to return to base and proceeded to do likewise. On landing I waited for some time and with some concern for my No. 2 to turn up.

When he finally appeared, and after some mutual discussion, it transpired that as dawn spread across the countryside the aircraft that he had been vainly pursuing at low level into occupied France was not my Spitfire but rather an Fw190 travelling very fast in a southerly direction.

Since he was alone and would have had no chance of closing on the Hun even if he had tried he very wisely decided to turn north and return to base.

In the absence of any further note in my logbook I can at this stage not be certain of the identity of my No. 2, which is probably just as well. Such a slightly ridiculous episode is probably best forgotten!

This rather quaint story is really the only memory I have of our very brief stay at Coolham.

Ken Lee, F/Lt NZ-39924
Auckland, New Zealand.

❐❐❐❐

Another pilot with 485 during its short stay at Coolham was Flight Lieutenant Des O'Halloran.

It was a long time ago, our lives were very eventful and, except in a few cases, details are beginning to fade.

To be honest, I cannot remember the airfield, the control caravan, yes.

After actuating the undercarriage during the process of landing there were two "bumps" indicating both legs had locked down. Not so, according to the instrument panel which displayed "Red Up".

After two attempts I informed the control tower. Our C/O, Johnny Niven, who was not flying on this sortie, climbed onto the roof of the control caravan to observe the position of my undercarriage.

After flying over at about 20 feet everything appeared to be in order. He gave his approval "it was my neck" to land with the undercarriage showing "Red Up".

The landing was normal.

Another incident which I am reasonably certain was at the "local" in Coolham concerns our arrival en masse which condensed the "regulars" into a corner. To us they all appeared to be very elderly and not very communicative. After accepting a "few" the thaw soon set in.

In return they took great pleasure in telling "their joke", which I am sure had been repeated over the years, to new ears.

This should confirm if my pub location is correct. Among the "regulars" was a gent with a well matured white beard. Around the walls were photographs, which I think were of the local horticultural club, showing the progressive growth of the beard in question.

Our very short stay at Coolham was a "practice" move, a 2nd TAF wing, under canvas, of three squadrons, each of 30 pilots, 20 aircraft, together with service and maintenance crews, required to relocate in 72 hours.

This was one of those moves in preparation for Normandy.

From Coolham we went to Funtington, then back to Selsey.

Des O'Halloran, F/Lt NZ-415418, married Mary, a Scot, who was a WAAF, in 1944. They live in Rotorua, New Zealand.

❐❐❐❐

Orderly Room Corporal Len Jordan recalls a flippant remark to a "rooky" that could have backfired.

Prior to the squadron's move to Coolham, armament had been restricted to .303 ammunition for the Brownings, 20mm cannon shells for the Hispanos and 250lb and 500lb bombs for the bombing attacks.

On arrival at Coolham the armourers came across other weaponry, namely rockets.

It was discovered that by carefully dismantling a rocket it was possible to use the propellant for experimental use as follows:

By packing the propellant into a circular cardboard case about the size of a cigar, attaching a piece of fuse wire and tying on a length of wood or stick, it was possible to produce for our own amusement a small sky rocket like a firework.

Not many of these, I am bound to say, were successfully launched.

At about that time a new "rooky" Aircrafthand, General Duties, was posted to the squadron. He was intrigued with these home made sky rockets and asked: "How do you get the stuff out?"

This brought the answer from F/Sgt Ernie Bonyard (a dour New Zealand armourer): "With a cold chisel and hammer," not expecting his answer to be taken seriously.

Shortly afterwards we heard the clanking of tools and realised the rooky had taken Ernie's answer seriously. Those around shouted to him to stop, he seemed oblivious to the danger.

That experience taught us all that in war one should not pass frivolous comments.

Len Jordan, Cpl 1874711
Northampton.

485 (New Zealand) Squadron
Funtington ALG, August 1944

Standing, from left: Cpl L.W. Jordan, P/O D.F. Clarke, F/O L.S.M White, F/O J. Yeatman, F/O A.B. Stead, P/O H.W.B. Patterson, F/Lt L.S. Black, S/Ldr J. Niven, F/Lt O.L. Hardy, F/O F. Transom, F/O T. Kearins, F/Lt D. O'Halloran, F/O D. Roberts, F/Lt K.J. MacDonald, F/Sgt C. McInnes, W/O M.A. Collett.

Kneeling: F/Sgt E.G. Atkins, W/O A.J. Downer, F/Lt W.A. Newenham, F/O R. De Tourret, F/O J.A. Houlton, F/Lt J.N. King, F/O A.C. Hutton, F/O D.G.L. Taylor, W/O J.J. Robinson, F/Lt L.M. Ralph, F/O M.C. Mayston, W/O R .M. Clarke.

Photo: L. Jordan

S/Ldr Johnnie Niven, DFC and Bar. *Photo: M. Collett*

15

2722 (Anti Aircraft), 2829 (Field) Squadrons, RAF Regiment

Leading Aircraftman Harry Matthews was a gunner with 2722 (Anti Aircraft) Squadron, RAF Regiment.

My first involvement with Coolham airfield was just before the buzzbombs started in 1944.

My squadron was in defence of it. My flight or platoon was spread out both sides of the Thakeham Road and the squadron's HQ was the large house south of the airfield.

The guns we manned were 20 millimetre Hispano canon and we were on duty two on, four off.

There were Polish squadrons there at that time.

My first engagement with buzzbombs was at night. The squadron moved out about a week after the buzzbombs started, to Folkstone.

Harry Matthews, LAC 1613860
Henfield, West Sussex

❑❑❑❑

Leading Aircraftman Jim Kennard was with 2829 (Field) Squadron, RAF Regiment and arrived at Coolham ALG on 17 April 1944.

In the early days, personnel were engaged in laying out camp roads and extensive training, including exercises using live ammunition, route marches of up to 25 miles and all aspects of ground warfare including, of course, airfield security.

A number of crash guards were mounted on crashed aircraft including one enemy plane.

As D-Day approached there was a marked increase in activity both in the air and on the ground.

I personally spent most of my days transporting 250lb bombs from Slindon Common which was a massive bomb dump.

On one trip we had a scare when a camouflage net slipped onto the propshaft between the cab and body of one of the vehicles in the convoy and caught fire! Fortunately, it was spotted and the fire quickly put out.

Often, between runs, we would help to bomb up the aircraft, one 250 pounder under each wing, to be dropped on gun emplacements or convoys and trains across the Channel.

The bombs were jettisoned if enemy fighters were encountered but I remember one occasion when a Spitfire landed with bombs still attached.

The pilot had shot down an Me109 but complained strongly to the mechanic that "ze plane, she is sluggish", etc, until the mechanic asked why he had not dropped the bombs. Was his face red, especially as it was contrary to orders to land bombed up!

Two things about Coolham which are strongest in my memories are a large pond stocked with big carp. The pilot of the plane I went up in on the Coolham Airfield Remembered weekend made a detour for me to see it from the air.

The other memory is of the Polish airmen putting everything into volleyball games as an energetic way of "relaxing" from the strain of cross-Channel sorties.

Jim Kennard, LAC 1217699
Dartford, Kent

2722 (Anti Aircraft) Squadron, RAF Regiment. *Photo: H Matthews*

16

422 Aviation Fuel and Ammunition Park Second Tactical Air Force

Every airfield needed competent back-up to set it up, keep it running smoothly and stocked up with the necessary equipment.

Leading Aircraftman John Raymond Whitfield said 422 moved to Coolham towards the end of March 1944.

The unit was made up of armourers, equipment assistants, motor transport drivers and aircraft hands general duties.

All our stores were delivered to us by road, the transport being three ton Chevrolet lorries.

The stores consisted of 1,000lb, 500lb and 250lb GP bombs, 20lb smoke bombs and 30lb flash bombs.

Cluster bombs were metal cases filled with 20lb fragmentation bombs complete with pistols, detonators and tail units for all bombs.

The ammunition was .303 (all types), .50, 20mm, 40mm, rocket projectiles, 60lbs rocket propellant which consisted of tubes filled with a solid cordite charge approximately three inches in diameter and five feet long. These were ignited electronically from the cockpit of the aircraft. Large amounts of aviation fuel were handled, 150 and 100 octane, in four and a half gallon jerry cans and the MT fuel was in four gallon cans.

Another item thrown in for good measure was a mobile oxygen unit. These were six ton lorries operated by airmen trained in the art of producing oxygen. This was compressed into large bottles for delivery to the airfields to be transferred to smaller bottles for use in the aircraft.

All the stores were loaded and unloaded by hand; we did not have the luxury of mechanical aids, that came later when we had a Coles Crane (mobile, of course) and a company of Royal Army Pioneer Corps.

For recreation we did the same as every other "bod" away from home – a glass in the local. I also used to look forward to sitting in a chair, one got fed up with sitting on bombs or "ammo" boxes.

Our sleeping arrangements left a lot to be desired as we were in tents most of the time.

We were not under the strict discipline of a fixed unit of the RAF. Maybe we were a "ragtail" mob but we did the job with a good heart. I, for one, am very proud to have served with 422 AFAP and the RAPC.

John Raymond Whitfield, LAC 1737645

LAC Whitfield driving a 422 A.F.A.P. Chevrolet lorry. *Photo: John Whitfield*

422 Aviation Fuel and Ammunition Park, Second Tactical Air Force. *Photo: Jim Whitfield*

17

The American Connection

By a strange coincidence the first and last planes to fly in to Coolham were American bombers which had to make emergency landings.

Both incidents were when the airfield was non operational, one when it was still under construction and the other after it was vacated.

Each of the crews of the B-26 Marauder, named Flounder Gus, and the B-24 Liberator, named Mad Madeline, had to quickly look for a place to land and Coolham's airstrip saved the day for them.

The Marauder, with five crew, had been "shot up" over Cherbourg on 15 February 1944.

The Liberator, with 10 crew, was returning from a bombing mission to Dresden on 26 January 1945, when it developed engine trouble, hit bad weather and could not make it back to its base in East Anglia.

Flounder Gus

1st Lt Richard "Dick" Lightfine was 21 and the captain of B-26 Marauder Flounder Gus, No. 41-34913, of 323rd Bombardment Group, 454th Bomb Squadron, 2nd Air Division, 9th Air Force, Earls Colne, Essex. He was on his 55th mission with Flounder Gus.

The "Crashed Aircraft Report" in official records states it landed at 11.00 hours at Farleys Farm at the edge of Coolham airfield and was "completely wrecked".

The report also records: One engine was out of action . . and as the plane was about to land the other engine cut out and the machine became uncontrollable. The pilot saw men working on the airfield and endeavoured to miss them, but failed to avoid Sgt 967357, Charles Gordon, of 5018 Airfield Construction Squadron, who was struck on the head and killed. The aircraft then tore off a wing and broke into two parts.

The crew of five were: Lt Richard H. Lightfine, Captain; Arthur Hanni, Bombardier-navigator; Staff Sgt Fay C. Kruger, engineer-gunner; Staff Sgt Rue Farnum, radio operator-gunner, and Staff Sgt Lloyd Raymond, tail gunner.

Farnum and Raymond were slightly injured. The entire crew was taken to the Canadian General Hospital, Horsham, where the two injured were detained. The three uninjured were later taken to RAF Dunsfold from where they returned to their base by air.

Dick Lightfine vividly recalls his emergency landing at Coolham and also reminisces about his Great Dane dog his Marauder bomber was named after.

We had been in London, Art Hanni and I, the night the first buzzbomb came over. We were late coming back from a mission so our regular hotel room was unavailable and we wound up in one of the Red Cross clubs and Art opened the window. There was a heavy fog and Art said: "Dick, look at this". I stuck my head out and all I could see was the glow of the after burner as the buzzbomb went by.

I thought nothing of it but the next morning the British were in panic because four bombs had come over that evening. One hit a marshalling yard, another a main warehouse, another a military installation and God knows what happened to the fourth, but they were panic stricken and rightly so.

From then on we were told to try and knock out the buzzbomb sites and that was our target in the Cherbourg Peninsular that day, 15 February 1944. But we got shot up quite badly. We had taken off at 8.45 in the morning and the chances are we had been ready two hours before having been briefed about 5.30–6am.

We had 12 planes – two flights – going to this particular target at Cherbourg. I was leading the 2nd Flight and probably nothing would have happened to me but for the fact that the lead crew got knocked out of the 1st flight so I sort of pulled up into position and as I did so I received a greeting from Adolph and his boys. The Germans were extremely good gunners and most accurate.

One engine was knocked out, the other engine was spewing oil like mad and I asked my engineer what should we do and he said: "Get home as quickly as we can 'cos I doubt we'll get there unless we start right now and the first field you see, land".

So we started back to England. The radio was shot out so we couldn't home in on Manston or one of the big bases and even if we had we couldn't have got there.

As we approached the coast I said: "Boys, you have a choice, you can bale out over water, you can bale out over land or we can take it in".

Art Hanni said: "I've baled out once, landed on the slate roof of an English mansion, bounced off that into oak trees and got beat up and under no circumstances am I going to bale out, we'll ride it in".

And Art spied a small grass airstrip so we came in. The right engine was feathered, the left engine still spewing oil. We circled the plane and we could see personnel on the ground watching us. I never realised that this was the first aircraft to land there.

We came in and just before we touched down, probably at 100 feet off the ground, the other engine conked out.

There was a tremendous yaw to the left and you could see them all running like hell. We did our best to straighten it out and Art pulled up the landing gear and we slid along the ground and hit a tree with the left wing. Unfortunately, a portion of it killed one of the young men on the ground. It was a shock to all of us.

The plane was completely demolished and broke in half but thank goodness it didn't catch alight.

But the tail section of the plane flew over a hedgerow at the end of the field and landed in a cottage yard.

In typically British fashion out walked a lady and here's this end of a fuselage lying there with a young American airman (gunner Lloyd A Raymond) and she said: "Would you like a cup of tea?" And he said: "Yes, I will have a cup of tea". So he proceeded to have a cup of tea and talk with her for about 15 minutes.

In the meantime we are getting our radio gunner, who had been knocked out, out of the plane and we started to look for Raymond and we couldn't find him.

All of a sudden he comes walking back, he'd been in a state of shock!

But someone came out and gave us some tea.

It was unfortunate that the engine didn't continue to function for another 15 seconds. Fifteen seconds more and we would have been down safe and sound and no one would have been injured.

It was thanks to Art pulling the wheels up to slow the plane down that more were not killed and also hitting the tree undoubtedly saved lots and lots of lives.

Afterwards I received a phone call and I was told: "Sir, the mother of the young man that was killed would like to talk to you".

It really shook me up to talk to her and find that her son had just come back from duty some place and she was so nice and said: "I understand it was not your fault, it was an accident and I want you to understand that I do not hold anything against you.

"It's an act of God and God bless you."

I had received superficial cuts to my head when we landed. The Canadians came over and took us to their hospital. Farnham was kept in with a broken nose and had been knocked out in the crash and gunner Raymond was kept in for examination because he was shaken up.

When the rest of us arrived back at base at Earls Colne the Flight Surgeon met us and gave me two bottles of Scotch saying one was for the enlisted personnel and one for the two officers.

He asked me if I wanted to spend the night in hospital and I said: "No, I'll sleep with the rest of the boys".

He gave me a couple of pills, I drank the Scotch with our friends, went to sleep and woke up next morning, quite late, and it was as if I had been drugged.

I saw the flight surgeon and asked him what on earth he had given me, I had never slept like that before in my life and he said: "Oh, it's something called twilight sleep. It's something that I give to my female patients when they're in pregnancy pain!"

Probably six or seven days later I was flying again.

Flounder Gus completed more missions without aborting than any other Marauder – a tribute to its dedicated ground crew who took care of that plane as if it was a member of their family. It was a magnificent plane.

Its replacement just didn't handle the same. It just wasn't such a good plane.

My co-pilot, before he took his own crew, had been Hank Sobczynski and he was the only other man to fly Flounder Gus.

After the crash Flounder Gus was taken to a graveyard some place and more or less forgotten until the air force sent down a large tractor trailer unit and brought the remains back up to our field. Whether they did this for a potential source of spare parts I don't know but I imagine that probably had something to do with it.

Then unbeknownst to me, at a later date, my friend, Hank Sobcznyski, had the section of the fuselage that had Flounder Gus's profile on it, cut out (and I didn't know this until many, many years afterwards) and he put it over his fireplace. He had also had a carrying container made so that when we went to the various reunions that the 454th had he had this along with him and we expect to bring this back to England in June 1944.

I never even knew the name of Coolham Airfield until contacted by Paul.

Regarding Flounder Gus, my magnificent Great Dane dog. He was given to me by my father and as a puppy he had never seen stairs. When he did he'd climb the stairs, his tail between his legs, his ears lying down flat and someone said: "Look at him floundering away".

So he became Flounder and then he became Flounder Gus and that was the name of our aircraft and we had a profile of him on the plane. He flew over to Europe with us but we never took him on a mission with us.

I once went to Simpsons restaurant in the Strand in London. I went in with him and tried to check him in with the hat check girl but in the end tied him to the serving trolley.

Three or four minutes later he fell asleep and he fell on his side, then his lead pulled the serving trolley into him and that caused him to start. The next thing we knew we have, right in the middle of the dining room, this Great Dane with the serving trolley behind him and yes, his tail was between his legs once again.

He could see everyone looking at him and when I got up to get him you could see the English looking at each other saying it was only a crazy American who would bring that dog in here.

When he went back to America with me I had to finish college and we did not have room for a Great Dane in the city who had never been on a lead. He lived the rest of his great life in the mid west with some people who could give him open space.

Dick Lightfine completed his quota of 72 missions, including one on D-Day and he recalls:

Looking through my log for 6 June 1944, it says: "Got up. Storm front over the peninsular".

And it's hard to realise how much action one saw on that day out in the Channel. All you could see was boats, ships, planes and the big shells from the battleships floating through the air.

I never thought I'd ever see anything like it in my life. I couldn't believe it.

On D-Day I only flew one mission because we were about the last ones to go out.

We were the last group to bomb on the actual German gun emplacements on Omaha beach on D-Day before the troops came in.

There were more missions flown later that day by different people but they had different targets.

Dick Lightfine was awarded the American DFC with two bronze Oak Leaf clusters, Air Medal with two silver and one bronze Oak Leaf cluster, the Purple Heart, Europe American Middle East Campaign with two bronze stars and World War II victory medal.

He lives near Chicago.

❐❐❐❐

Lt Arthur Hanni was the bombardier-navigator and emergency co-pilot on Flounder Gus. He recalls their mission to Cherbourg was to bomb a rocket launching installation and these are his memories of their emergency landing at Coolham Airfield and he too, had not known the exact location of the crash until contacted by Paul.

After bombing the target and turning away for home our right engine was hit and knocked out by an anti-aircraft flak shell.

Lightfine finally got the prop feathered and we dropped to an altitude of about 5,000 feet from about 12,000 feet and naturally falling behind our group because of loss of air speed.

I moved from the bombardier's nose section to the co-pilot's seat. Approaching the Channel we thought we might have to ditch in the Channel so crew members threw out anything of weight that we couldn't use – flak vest, ammunition, guns, etc.

Our left engine was running about full blast to maintain minimum altitude and necessary air speed. After a seemingly long time we reached England and desperately looked for a base to land.

The visibility was bad and for a time we couldn't see a place to land.

Finally I saw what looked like a grassy landing strip in a field. I pointed out this location to the pilot and he started to descend and line up with the landing strip.

At about 100 feet the left engine conked out and with the trim controls rolled in and because of the right propeller windmilling the pilot could not control the plane and we started veering sharply to the left.

I yanked up the landing gear for a belly landing to slow our speed.

We were heading directly for a big tree and hit it, sheared off the left wing next to the pilot, hit the ground skidding.

The tail section broke off with the tail gunner and it soared over a road and hedge and landed very near a farmhouse. He crawled out, badly bruised, with minor cuts.

The radio man was knocked unconscious with a broken nose. The engineer and I picked him up and removed him from the plane.

Shortly thereafter we were taken to a hospital and sometime later flown to our base at Earls Colne.

After recovering we flew more combat missions.

The crash and the crew's escape from it, was reported in Arthur Hanni's local newspaper, *The Topeka Journal,* at the time. Its details had been released in a '"dispatch from a bomber base of the Eighth Army Air Force in England".

The report gave graphic details and referred to the tail section soaring over a hedge and road carrying the gunner, Staff Sgt Lloyd A. Raymond, and to the rest of the crew rescuing radio-operator gunner, Staff Sgt Rue Farnum, knocked unconscious in the crash.

"After giving him a first-aid treatment they carried him to the farmhouse where they found Sgt Raymond badly bruised and cut about the face.

"He had been able to climb from the tail section and limp to the building under his own power. Meanwhile, the farmer's wife was boiling water to make the boys a cup of tea."

Lt Hanni was promoted to Captain and completed a total of 70 missions. He was awarded the American DFC, Air Medal, three Oak Leaf Clusters and the Purple Heart for wounds received in air action over England.

He lives in Kansas.

❐❐❐❐

Staff Sergeant Rue Farnum, radio operator on Flounder Gus, has "vague" memories of the crash.

We were stationed at Earls Colne for 15 months. On one of our missions over France we received damage to our plane, and returning from this mission we made a crash landing at Farleys Farm, near Coolham.

My memory of that incident is very vague as I was knocked unconscious.

That was my 34th mission and continuing to fly missions, I completed 64.

Rue Farnum
Fairbank, Iowa.

Crew of B-26 Marauder Flounder Gus
Photograph taken March 1943, Barkside Field, Louisiana, USA

From left: Staff Sgt Lucien Thomas, Lt Richard Lightfine and his Great Dane Flounder Gus, Lt Arthur Hanni, Staff Sgt Rue Farnum, Staff Sgt Fay Kruger.
Photo: A Hanni

Playtime for Great Dane Flounder Gus posing with his master, Lieutenant Richard Lightfine.
Photo: R. Lightfine

B-26 Marauder Flounder Gus. *Photo: A. Hanni*

Remains of B-26 Marauder Flounder Gus after its fateful crash at Coolham ALG whilst completing its 55th mission on 15 February 1944. *Photo: A. Hanni*

Mad Madeline

Russ Journigan was the Bombardier-Navigator on the B-24 Liberator Mad Madeline which made an emergency landing at Coolham on 19 January 1945. It had beem on an 11 hour mission with 10 crew on board and Russ was aged 20.

We were attached to the 491st Bomb Group, 855th Squadron, 2nd Air Division and our aerodrome was located at North Pickenham in East Anglia.

We were flying with a senior pilot, Lt Frank Goodrich, whom I have not been able to locate.

We were returning from a bombing mission to Dresden in eastern Germany when we experienced turbo supercharger trouble after the bomb release and had to drop out of formation. The reduced speed put us behind in returning back to the UK.

Due to extremely bad weather, with snow and a strong cold front, we could not make our let-down over East Anglia. I gave a new heading to Lt Goodrich to go south to get out of the storm.

It was 10-10th cloud cover and we had not seen any part of England on our return from the Continent.

We were running low on fuel after being on an eleven hour mission. We called "Darkie" over southern England, an emergency aircraft frequency that all aerodromes monitor. There were no responses.

I was doing dead reckoning on a Mercator chart and using the RAF "G" Box that we had installed in the plane to get our positions.

I didn't want to go too far south in England and run out of any place to land and possibly have to ditch for lack of fuel.

I was located in the nose and just by luck, saw an opening in the cloud cover.

I notified Lt Goodrich on the intercom, who had also noticed the hole.

This was the first time we had seen any land at all on our return. We made a quick descent to get under the cloud cover. There was about a 400 foot ceiling.

As we were short of fuel everybody in the plane was looking for a place to land, when ahead of us we located Coolham.

Quickly dropping wheels and flaps we lined up on the north-south runway. We didn't even circle the strip.

It was raining with some sleet to reduce our visibility and just keeping the ship lined up on the runway was difficult.

I don't remember there being a wind "T" or a sock to determine wind direction. We had no chance to circle to find out. We just dropped the big bird in!

I had hoped that the wind was coming from the north somewhere, but upon landing there was a crosswind which didn't help the landing.

With everyone on the crew who could see, on the intercom helping the pilots, we touched down, and you might say, we were happy to get down.

After touchdown the plane veered off the metal strip and into the mud which helped us stop.

After getting out of the plane I remember that mud was splashed all over the bottom of the plane and some mud even got as high as the waist ports where the .50 caliber machine guns were mounted.

After about two minutes, when we disembarked the bomber, we were met by about eight English children, who rolled up to our plane on bicycles. They had met 10 tired, hungry warriors, happy to be back on the ground, with rousing cheers of "smashing good landing, smashing good landing", an expression that we Yanks didn't quite understand but later found out meant "good" or "excellent"!

We still didn't know where we had landed since Coolham was not indicated as an airfield on any of my maps that I had on board. We didn't even get any headings to any of the aerodromes in southern England that we would've gotten had any of our "Darkie" distress signals been answered. No one had returned our frantic calls.

After being on the ground and surveying the damage, we inquired about the location of the nearest aerodrome which was the RAF base at Dunsfold (in Surrey).

As I recall, a number of cars showed up and the crew was taken to the manor house in Coolham where the English people in the village provided enough rationed food to feed us dinner there.

All of us really felt good about the villagers' hospitality, especially to American airmen that they were not accustomed to seeing, southern England being so far south from Eight Air Force operations, concentrated mostly in East Anglia.

I do remember the manor house being very old as you had to stoop down to go through the doors.

After notifying Dunsfold Aerodrome of our landing a lorry was sent to Coolham to pick some of the crew up. Myself, Lt McPhillips, the co-pilot and all of the enlisted men, except the Flight Engineer, Tom Eaton, who stayed with the airplane along with Lt Flaherty and Lt Goodrich, were trucked back to Dunsfold that evening.

About half way back we stopped at a pub and in walked seven Yank airmen in full flying gear to get a beer. Needless to say, the local people probably thought they were being invaded by some people from outer space as our oxygen masks were still hanging from our helmets and we were in electrically heated flying suits and fleece lined boots.

After a beer, we got more friendly and at that time I learned how to shoot darts and play a game call "shove-a-penny" for the first time.

As I remember, while riding up front in the lorry, the pub was located on the straight part of the road going to Dunsfold on the right hand side of the road, going north.

We arrived at Dunsfold Aerodrome in the middle of the night where I met the Orderly Officer at the main gate. We were assigned to our billets, one for the enlisted men and one for me and Mac in separate quarters.

It wasn't until the following morning that we discovered that we were billeted on the WAAF site for the aerodrome.

An interesting thing happened that morning when some

of my enlisted men went down to an ablution, which was common to the site, to wash up, when they discovered women taking showers and in various states of undress!

The Orderly Officer had not informed any of us the night before that we would be in a co-habiting living mode! Anyway, everything worked out to everybody's satisfaction.

I do remember running out of coal for the small space heater located in my billet. Early the following morning I went outside in the cold to pick up some coal. I located some scattered on the ground and picked up what I thought was coal.

To put this incident in the proper perspective, I came from California and had never had to use coal for heating as we used natural gas for heating.

When I was asked by a WAAF what I was doing I replied that I was "picking up some coal for the space heater". She said: "That stuff won't burn, they are clinkers or burnt out coal," and that I would need to go down and get a bucket of coal at the collier. I was embarrassed by this chain of events.

I was the liaison officer between the operation at Coolham and our aerodrome at North Pickenham. Since we were flying maximum efforts during the Battle of the Bulge our Commanding Officer needed to know when we could expect to get our airplane back into service.

Because of the short strip at Coolham we decided that in order to get the airplane off the strip we needed to lighten the ship as much as possible.

We took all the armament and ammunition out of the plane along with the armour plate that could be unbolted.

As I remember, we had only about 400 gallons of fuel left in the wing tanks. Lt Goodrich determined that we needed only enough fuel to take off and land at Dunsfold, the closest base where we could refuel. We pumped out about 200 gallons to lighten the load.

One of the major problems was in getting the ship out of the mud and back on the metal mesh runway. We enlisted the aid of a small farm tractor from a farm nearby and by attaching a chain to the nose wheel and the main landing gear we tried to move the bomber inch by inch.

Because of the beastly cold weather with snow, rain and ice, we did better in the early morning when the tires could get up on top of the frozen ground by using planks. When the ground would warm up the wheels would sink back down in the mud again by afternoon. This really slowed down the operation.

As I remember, we burned the clutch out of the farm tractor and I sincerely hope that the farmer was paid for a new clutch.

The Deputy Group Commander at our aerodrome was a Lt Col Jack Merrill and he was getting impatient on our progress. I got the feeling that he thought that we were just fooling around when actually, we were making every effort to free the plane from the muck, under adverse weather conditions and with no big equipment.

I can still hear the Colonel say: "Journigan, when are you going to get that plane back here?" where I would promptly say, "Sir, tomorrow for sure".

We were down there eight days trying to get the plane in a condition and position to take off again.

Meanwhile, at Dunsfold Aerodrome, I went down to the Control Tower where I met F/Lt Pierce, an RCAF Flight Control Officer for the aerodrome.

I requested some different air maps which would show the Coolham airstrip. He finally located one for my navigator's flight log case.

We had a terrible snow storm on about 22 January 1945, at both Coolham and Dunsfold and everything came to a standstill.

As I can best recall, Dunsfold was flying Typoons and Hurricane fighters. The Commanding Officer at the aerodrome ordered that personnel be used to clear the runway of snow by hand shovels.

I dispatched myself and my enlisted men to help clear the runway of snow. I had thought that since we were eating their food that we should help with the chore.

Much to our surprise, about 1pm, a tea wagon approached on the runway and the work stopped and everyone had tea. It was the first time that I had ever had tea with milk in it. It really felt good inside, with all that cold weather, in spite of us still being in our heavy flying gear.

Later that day, some enterprising RAF chap suggested that if we could line up a fighter plane in the right direction and fire up the engine that the prop wash would blow the snow right off the runway.

It worked. In no time that job was done. Looking back it would be nice to meet that forward vision chap who shortened our workday by about six hours, as they were planning to work into the night to get the job done.

I accompanied F/Lt Pierce back to the Officers' Club for refreshments where I was introduced to a game called "Bottle". I had never played this game before but, of course, the RAF officers were quite good at it!

The game is played with a number of people forming a circle. Each person counts a number, in succession around the circle. When you get to a seven, ie, 7, 17, 27, etc, or any multiple of seven, ie, 14, 21, 28, 35, etc, instead of saying that number you say "bottle" and the counting rotation reverses.

The rub comes when your turn comes and you fail to respond correctly. If you miss, you buy everybody playing a round of drinks.

As you might have guessed, the young Yank lieutenant of 20 years was one who made most of the mistakes. As the evening wore on, my mental facilities were not quite as alert, which didn't help things either!

A good time was had by all even though I felt a little disadvantaged having not ever played the game before.

I had the pleasure of meeting Wing Commander Johnnie Johnson, a fighter ace, who was visiting the aerodrome. I didn't know at the time what a hero he really was until I read the history about him later in my life.

Saturday night rolled around at Dunsfold and a Leading Aircraftwoman, WAAF, invited me to the dance.

Her name was Gwendolyn Young, another RCAF person whom I had met at the Control Tower.

Now, the problem was I was still wearing my smelly cold weather flying gear and I was going to the dance with this nice blonde young lady. I told her that I was not properly dressed for such an event but she assured me that my dress was Okay and insisted that I attended the dance.

I remember that we went to a community hall where a gramophone played the dance records with some trumpet loud speaker horns coming out each side of the control amplifier and a man drummer kept the beat to the record machine. Looking back, the first real Hi-Fi!

I was not used to dancing in a circle as most of my dancing with the "big bands" in California was helter-skelter in direction.

Can you imagine what I must have looked like with all these nicely dressed Wing Commanders, Squadron Leaders and Group Captains and me "cutting a rug" in my smelly, dirty, heavy, flying suit with clumsy fleece lined boots that you could hardly walk in!

What a sight it must have been, especially with this beautiful blonde from Calgary, Canada, who had broken a date with an RAF officer to go to the dance with me!!

Meanwhile, back at Coolham, all the extra weight was taken out of the bomber and shipped to Dunsfold by lorry except the extra fuel which was left at Coolham.

The plane was finally positioned for take off. To take advantage of every inch of runway, the tail stuck back over the hedgerow that formed the out perimeter of the landing strip. All four engines were started quickly and brought up to full power. With the ship shaking every rivet, Lt Goodrich and one Flight Engineer released the brakes and the awkward ship started to accelerate down the metal meshed strip.

At lift-off, not much room appeared between the bottom of the plane and the hedgerow at the other end of the field but the ship got airborne.

I jumped in the lorry and the driver drove us back to the Dunsfold Aerodrome where happily we found the ship parked on the tarmac in front of the Control Tower.

Anxious to get back to our base, the remaining crew loaded up all the stuff we had taken out of the ship at Coolham.

We gassed up for our journey to North Pickenham. The weather was still frightful and it was necessary to make another emergency landing at Hethel Aerodrome, about 40 miles from our aerodrome in East Anglia.

Again, the snow was so bad we had to stay another night at Hethel.

The next morning we got enough snow off the wings by spraying alcohol on the wing surfaces and brushing the snow off.

We took off once again for our destination and, pleased to say, we made it back this time.

Everyone was glad to see us back including my cocker spaniel who probably wondered what had happened to me for eight days.

Other officers in my squadron tried to feed him but he just lay on the bottom of the bunk and didn't eat.

I went on to complete 22 missions, being transferred to a Lead Crew as a Lead Bombardier.

After VE Day we flew our "war weary" ship back to the States on 22 June 1945.

All the aircrew members were given a 30-day leave to go home and returned to duty at McCord Field in the State of Washington for transitional training on B-29 bombers for the war in the Pacific.

In August 1945, the war was ended with Japan and I went into inactive duty with the reserves. The plane was ferried to Davis Monthan-Field in Arizona where most of the "War Wearies" in the Eight Air Force ended in the scrap heap.

But I will always remember this unforgettable experience in attempting to make an emergency landing with 24 tons of metal and one ton of human flesh, trying to figure out in not much time how to put this big "gas bird" on what appeared at the time to be a small path in the south English countryside called Coolham. God must have been with us on that trip!

Russ Journigan, (Lt USAAF Ret), "Ex 8th Airforce bomber flyer who was lucky enough to get back".

He lives in Hawaii.

Lt Russ Journigan

Crew of B-24J Liberator, Mad Madeline

Standing, from left: Tech Sgt Frank Eaton, Lt Norman Flaherty, Lt Russell Journigan, Lt Paul Church, Lt Frank McPhillips, Sgt Frank Eckert.
Kneeling: Sgt Robert Mathews, Sgt Bill Ellington, Sgt Robert Lease. *Photo: R. Journigan*

B-24J Liberator Mad Madeline, foreground, in formation with B-24 Liberator Big'un. *Photo: R. Journigan*

Aerial view of Coolham ALG, 16 April 1946.

Crown copyright/MOD

18

Memories of Coolham – Civilian Workers and Villagers

Coolham Advanced Landing Ground airfield had an enormous impact on the lives of those who lived around it.

From its earliest beginnings, to when it was operational, followed by the eerie silence when the squadrons were pulled out, people still have vivid memories of those exciting, dangerous times.

Many come from those who were children during the war who typcially seemed to have no sense of danger and some of their recollections make your hair stand on end.

One of the first to be involved with the airfield was Joy Coleman's husband, Ronald, a civil engineer employed by the then Ministry of Transport, who was seconded to the Ministry of Defence.

He had a yard in Roffey, Horsham, and he and his "gang" carried out work on the ministry's behalf such as hard standings for heavy vehicles, ammunition dumps, etc, with an eye to future use when the Second Front was opened.

Joy recalls that one afternoon in early spring she was working in her garden at their home in Horley, Surrey.

It was a Sunday, but my husband, like so many others at that time, seemed to work all the hours of daylight, seven days a week.

A tall stranger in RAF uniform appeared and asked for my husband. I explained that he was not available and probably would not be for some hours.

He then opened his briefcase and took out a bulky packet, sealed with tape and bearing a "Secret" sticker of the kind I had grown familiar with when plans came from the Ministry of Defence.

"Put this away safely and give it to your husband as soon as he comes home," he said. "He will know what it is."

I asked who should I say had called, he hesitated a moment and then replied: "Squadron Leader Lancaster".

I was rather amused. Nine out of ten of the bombers setting out from this area had been Lancasters so his choice of a nom de guerre was rather obvious. (It turned out to be his real name, O/C 5004 Airfield Construction Squadron.)

My husband left the next morning to go to Coolham first.

That was the beginning of turning some acres of farmland, dotted with beautiful mature oak trees, into Coolham Advanced Landing Ground.

An old manor house, with its lovely garden and greenhouses also had to be sacrificed though my husband tried hard to save them.

He employed a local contractor, working with some of his own men.

Once the project was well on the way to completion he was sent off to do something else, leaving his assistant in charge, so he never saw the airstrip operational.

I had forgotten Coolham until my daughter, now living within a few miles of the site, mentioned it.

❑❑❑❑

Brigadier Lionel Lees, FRCOG L/RAMC, lived with his elder brother, Stephen, aged 13, at Jendens Farm, Shipley.

It was Spring 1943 and we watched in awe as the massive bulldozers charged into centuries-old oak trees and pushed them over as if they were mere saplings.

For months, surveyors had been tramping over our fields with red and white poles and peering through theodolites and we never really believed the terrain of the heavily wooded countryside could possibly be as flat as they tried to convince us. When the trees were gone, however, we could see that the earth-movers had relatively little to do to level the ground.

Each day we rode on horseback around the commandeered land and watched as the airfield rapidly took shape. A perimeter track for service vehicles was excavated all round and filled with hard core and a strip down the centre was overlaid with a coarse wire mesh known as Sommerfeld tracking.

The runways and the unmeshed area on either side were then seeded with grass and in a very short space of time the land transformed into a huge lawn.

During this time a Sergeant, whom we knew quite well by sight, was killed while supervising a track laying party. A crippled Marauder bomber was forced to make a crash landing on the uncompleted runway and the unfortunate Sgt Gordon was struck by a propeller blade.

By Spring 1944 the two-runway airfield was ready for use. A number of large blister hangars and some smaller buildings had been constructed, aircraft dispersal areas had been set out around the periphery and, in a field a mere 100 yards from our house, piles of 500 pound bombs and five-gallon flimsy cans of aviation spirit lay ready for use.

At last the aircraft arrived and we watched excitedly as the squadrons of Mustang fighters circled in formation, spread out in line astern and landed one by one.

We noted the different squadron letterings, which we were to learn by heart in the months to come and we stayed until the last 'plane had taxied to its allotted dispersal point and the whirring propeller blades clattered to rest.

As there was no perimeter fence around the airfield, we were free to wander at will among the parked aircraft, chatting to the Polish airmen, many of whom had scant understanding of English, and clambering into the cockpits to play with the joystick and stretch to reach the rudder control pedals.

The groundcrew were remarkably tolerant, answering our endless questions and allowing us to sit on the bomb trolleys to write messages in chalk on the bombs such as "Present for Adolf", and "Goering, this one is for you".

However, we overstepped the mark on one occasion, when we went to stand beside the aircraft controller, whose task was to signal with red and green lights to incoming aircraft and warn them with red Very lights if they had failed to lower their landing gear.

We were thrilled to have the aircraft passing over us a bare 20 feet above our heads, but clearly our presence was a distraction to the controller, and we were marched off by an irate duty officer.

Our usual vantage point was a large mound of topsoil which had been bulldozed to the south side of the east-west arm of the airfield, just one field away from our house.

We shared our vantage point with members of the RAF Regiment manning an airfield defence 20mm cannon and from there we had a grandstand view of all the flying activity. We recorded each aircraft as it took off for a sortie over France and duly noted their times of return.

Soon, whole notebooks were filled with these jottings, the continuity of the records being interrupted only by the inconvenient hours we were obliged to spend at school.

Occasionally, there was sadness when an aircraft failed to return and one day we watched in horror as a Mustang overshot the north-south runway, skimmed the lake at Coolham House and crashed into a line of poplar trees beside the Coolham Road, exploding into a fiery inferno.

There were more lighthearted incidents also, such as the day we dived into a ditch as bombs beneath the wings of a Mustang became detached shortly after it left the ground and came bouncing and rolling towards us.

Or the occasion a taxiing Auster light aircraft was caught by a crosswind and turned gently over onto its back, leaving the pilot to crawl out unaided, shaken but unhurt.

During the Spring and early Summer of 1944 the grass on the airfield grew rapidly and my father was given the contract to keep it mown. He was allowed to remove the cut grass as hay for the farm livestock and this gave rise to one of the lighter episodes of the year, though it could have had serious consequences.

Jim Johnson, our elderly farm worker of pure Sussex stock, had been despatched to the airfield with his horse and wagon to pick up a load of hay.

The wagon was but half full when a squadron of Mustangs returned from a sortie and wished to land.

Jim, however, refused to move until his load was complete and felt deeply affronted when he was escorted from the runway under armed guard.

Another of our perks was the contract to collect pig swill from the airbase and I used to accompany my cousin Peter, driving our ancient Humber farm car and trailer round the perimeter track, calling in at each squadron cookhouse.

The car went beautifully on aviation spirit until eventually the high octance fuel burnt out the valves!

Three months after their arrival, the Mustang squadrons moved on. They were replaced by three squadrons of Spitfires, but these stayed for only one week, then the airfield was operational no more.

An eerie stillness descended on the wide open spaces where recently there had been so much noise and activity.

We had become so accustomed to the need to raise our voices to make ourselves heard above the roar of engines, that now we felt we should speak in hushed voices so as not to shatter the almost tangible silence that surrounded us.

Though decommissioned, the airfield was not returned to us to use as farmland for another year.

During this time one echo of its glory days returned as a reminder that it had not all been a dream.

In January 1945, a B-24 Liberator bomber made an unannounced emergency landing, coming to rest off the Sommerfeld matting at the Coolham end of the runway, where it sank some two feet deep into the wealden clay.

The 10-man American crew were thus marooned on a deserted airfield while they waited for it to be heaved out.

During their stay, members of the crew came to our farmhouse in relays for food and hot baths, always leaving some of their number to guard the aircraft. In return they emptied their emergency rations to give us chocolate, nuts and dried bananas, which were luxuries indeed for Stephen and me.

When it was ready to be heaved out no amount of engine revving and digging around the wheels would liberate the Liberator from the fastness of the glutinous clay.

My father lent his Fordson tractor to assist, but its spade-lugged wheels merely dug deep gouges in the soft ground and eventually its clutch burnt out.

After a week, a very muddy bomber was at last winched out by a heavy aircraft recovery vehicle and, after a final low pass over the airfield, disappeared into a leaden Winter sky.

The Sommerfeld mesh was finally ripped up some nine months later, leaving, as we were to discover, many of the plough-breaking steel pickets behind.

The loss of so much valuable farm land for a period of three years, and the work involved in getting it back into production again, cost my father dearly but, as a childhood memory, it was a time of excitement and adventure.

❑❑❑❑

The grass cutting at the airfield, referred to by Brig Lees, was carried out by Donald Mitchell.

Mr Lees, who then farmed Jendens Farm, was asked to mow the runways and around flags and bomb stores but sub-contracted the job to Mr Tom Thornton, of New Barn Farm, Dial Post, who I worked for.

The price, I think, was 10 shillings per acre based on 120 acres.

Sometime in the summer of 1944 I spent three weeks at Coolham aerodrome with a tractor and a six-foot-cut Bamford mower. The tractor was American brought over on lease lend.

Due to the nature of the grass and the fact I was only allowed to work when there were no planes flying out or home the job took longer than anticipated.

I think flying was approximately between 8am and 8pm and because of double summertime I often worked until 11.30pm and then cycled home to Normans Cottage, Ashington, and usually returned in time to do some mowing before flying started.

❑❑❑❑

Tony Overington, of Newpound, Wisborough Green, spent a great deal of his time during the war with his grandparents, the Patchings, at Milton Cottage, Coolham, opposite the old school in Thakeham Road.

I can remember the ground being cleared for the aerodrome.

They hauled all the large oak trees to J. & S. Agate (timber merchants in Horsham) with two steam traction engines.

One of my pals was Charles Creasey, who also lived with his grandparents at Foxcroft Cottage, Thakeham Road, which backed onto the western end of the runway of the aerodrome.

Prior to the building of the drome, as it was affectionately known, there were four very large horse chestnut trees along the back boundary of Foxcroft Cottage and considerably taller than the house.

These were felled, for obvious reasons, but the stumps left were about four feet high and made very good vantage points for two inquisitive schoolboys.

It was from here that Charles and I witnessed some great flying feats by pilots bringing in shot-up planes, some with tragic results, one which Charles and I were involved in.

We always knew when there was going to be an emergency landing because the fire and rescue tenders would come speeding down the runway and station themselves at the back fence of Foxcroft and about 20 feet or so from our chestnut vantage points.

One particular evening, in June 1944, having seen the rescue services take up their positions, Charles and I climbed onto ours. Very soon we saw two larger planes than we were used to, both twin engine jobs, coming in from the Shipley end and one had its port engine on fire. The other was acting as an escort.

They were American Lightning P-38 fighters, unusual in that they had twin fuselages. The damaged plane landed and came careering down the runway towards us and virtually out of control.

The second, the escort plane, realising it was unsafe to land, lifted its undercarriage to do another circuit.

He barely missed the chimney pots of Foxcroft. As he flew over us we both turned our backs on the runway to watch him and the next thing we knew Charles' grandfather, Harry, who, incidentally, was stone deaf from the 1914/18 gunfire, snatched us both from our vantage points to the ground as the wing of the damaged plane came over the garden about two feet above our tree stumps, taking the washing off the line and the heads off Harry's sunflowers.

Had it not been for Harry's quick thinking I would not be telling you about it. Apparently, one of the wheels of the Lightning had dropped into the safety ditch which was about 10 feet from the hedge and it slewed the Lightning round almost facing back the way it had come.

It burst into flames seconds after they got the pilot out who had been wounded in the leg.

We stood by the ambulance as the medics picked him up and he said that he could see us both there, with our backs to him, and there was nothing he could do about it. I have often wondered what became of him.

I remember Coolham ALG as a very cosmopolitan station with its multi national aircrews.

If the pilots had been successful in their missions the Polish chaps, who were right tear-aways, would come in low, while the Brits would stick to the more conventional aerobatics, some of which were quite spectacular.

One other amusing point, just along the road from Creasey's lived an old chap called Jim Venn, a great man for his ale. He used to leave the Selsey Arms very much the worse for drink and when he got to Creasey's he would get off his bike and push it the rest of the way home.

When we asked him why he did this he said because one night one of those low flying planes had blown him off his bike and he was not going to let that happen again.

But I would venture to suggest that it was the ale that

was the cause of the trouble, not the low flying planes!

One of the most vivid things that became ingrained into my young mind was the sheer hatred which these Polish chaps had for the race which had over-run their country and who they were now fighting from Coolham.

I was asked one day if I would like to go for a ride to the blacksmith's at Shipley to watch him put an iron tyre on a wheel repaired at Hardings, the local builders/ wagon builders' yard.

The blacksmith's shop at Shipley used to be almost next door to the Blacksmith's Arms, now the Countryman Inn.

At one point, as we approached Old House, we saw a lot of airmen and soldiers about.

As we reached the old electric light house we saw a string of ammunition and bombs scattered along the road.

I picked up some aerial photographs, all of which had scorched and burnt edges.

The airmen told us that a Mustang had just crashed but would not say where. When we got back to Coolham I showed the photographs to Mr Harding, who was a Special Constable, and he said we should hand them in.

Some days later he told me they were photos of the damage our bombers had caused to the Rhur in Germany.

A coincidence that occurred after the war again revived my memories of Coolham airfield.

The station commander at Coolham was a Squadron Leader Haw whose nickname was Wag.

Some years later, in the early fifties, I was a young apprenctice carpenter working on the restoration of the Blacksmith's Arms at Adversane.

The landlord was S/Ldr Haw who, incidentally, was married to the Craven A girl. She appeared in the cigarette adverts wearing a big floppy hat and a matching polka dot dress.

They had a son, Stuart. S/Ldr Haw asked me if I would carve his son's initials into a new oak beam that I was putting into the ceiling. This I did and S/H can still be seen there today.

❑❑❑❑

A teacher at Rydon School, Thakeham, during the war, Phyl Humphrey, vividly remembers D-Day, 6 June 1944.

I was awakened at dawn by the drone of many engines as trains of gliders of airborne troops were hauled overhead towards France.

I also remember during that summer hearing a very loud crash in the vicinity of Old House, now St Julian's, on the A272 road near the bridge, east of the village.

I dashed down the road on my cycle towards a small group of locals and Polish airmen surveying a broken topped conifer. Lying in the road was a very large unexploded bomb. An aircraft must have taken off and topped the trees causing the bomb to have been raked off.

No one seemed particularly worried that the unexploded bomb may have been in an unstable state and the incident was soon forgotten.

As well as the airmen stationed at Coolham there were many more British and Canadian troops in great concentration in the area. Philip Hales, a local business-man who also farmed at Brooks Green, and his wife, did much to supply some light relief for the young pilots and key personnel from the training camps. He held Saturday night parties in a large playroom at the end of his garden and while our elders played bridge, we danced to the radiogram.

On Sunday afternoons, weather permitting, we played croquet and had tea on the lawn.

On one of these occasions one young pilot nicknamed Geordie (P/O George Pyle of 129 Squadron), who had been specifically invited, failed to turn up. His fellows glumly told us that he had failed that day to return from a hedge-hopping flight over occupied France.

The news had to be broken by our hostess to his girlfriend. We had no news for some time as to whether he had survived but many months later, our host having made inquiries, he turned up at a party put on in his honour at the farm.

He had been shot down and a French farmer had helped him burn his plane and had hidden him until the Resistance could smuggle him out of the country through one of their many escape routes and he returned to England.

He was not allowed to give details, for obvious reasons, but he did show us a snap of himself in a rather shabby sports coat with blue beret, masquerading as a French-man, with very little French at his command.

He also showed us the printed neck scarf worn by pilots which had printed coloured maps of the enemy terrain on them.

Dances were also put on for the troops by two members of the Women's Institute, Mrs Creasey and Mrs Humphrey, in the village hall.

❑❑❑❑

Wally Standard was a 17 year old agricultural student in 1944 whose parents lived in The Nook, next to the village hall.

On one of my visits home the village was alive with air force personnel and the activity of the aircraft was most exciting.

There were normally two main raids each day, compris-ing 36 aircraft on each, one am and one pm.

I came to know a few pilots by name in the international squadron (129) and I particularly remember the Austra-lian, Roggie (Neil Roggenkamp), who had his name in white across the back of his flying jacket.

The only contact I remember having with a Polish airman, who was not a pilot, was through his English wife and little girl who stayed at our home from time to time during the operational period.

Delayed returning to college after I broke my wrist I used to stand on the B2139 road to watch the departure and return of the aircraft.

One day, when the wind was from the north and I was at Sprouts Corner watching the aircraft return from their morning mission, which were stream landing from the south, one Mustang approached without its landing wheels being lowered and the airman who was standing at the end of the runway fixed two verylights.

The pilot took no notice and landed with wheels up, tearing up the runway and damaging the aircraft. Chaos reigned for a few minutes as the other pilots waiting to land sorted themselves out and finally approached the east/west runway from the same direction. All landed safely.

The problem arose during the afternoon when the second sortie of the day was to take off crosswind, in pairs. As Roggie turned on to the runway he waved to me and moments later I was horrified to see his aircraft swing to port as its tail lifted before take off.

He did not leave the ground but crashed through the eastern end of what was left of a wooded area. A further aircraft swung similarly on take off, got off the ground but stalled, falling into the wooded area.

To my surprise I saw Roggie the same evening, he was unhurt and said that he had never left an aircraft so quickly. The other pilot was also unhurt.

My mother was caretaker of the village hall throughout the war and helped others organise events including the dances. She frequently operated the radiogram, on permanent loan from a chicken farmer from Brooks Green, using the old 78's.

My other memories include seeing doodlebugs flying over and saying the usual prayers as they approached, and also the Liberator stranded on the airfield.

❑❑❑❑

Barry Lenharth was a schoolboy aged around 10 when he and his pals would cycle to the airfield from West Chiltington Common.

Our first visits were confined to peeping through the hedge. Gradually we befriended the airfield personnel and were allowed on to the airfield itself getting to know many of the ground staff.

I became a particular friend of Michael Tanner-Garbarz (of 306 Squadron). One day he asked me if I knew the words to the song, "When you're smiling".

At the time my parents ran a concert party entertaining the troops. When I asked them to write the words down they invited him back to our home for the evening.

This was the start of a long-term family friendship and in later years when Michael returned from Belgium with his wife they came to live with us for a while.

The airfield seemed like a great adventure to us. We could only see the glamour and, as boys, did not consider the harsh realities of war it represented.

We used to watch the planes and armaments being serviced and repaired. I recall that sometimes we were even able to sit in the cockpit and look through the gunsight.

On occasions they would even fire up the engines while we were seated there and memories of the tremendous roar and vibration remain very clear to this day.

At meal times we were looked after very well. I remember sitting on big wooden benches with the sides of the canvas mess tent rolled up during the summer.

We were given great mugs of sweet tea, enormous bacon sandwices and hunks of cheese, bread and butter. In the sparse war days, this was easily a week's ration!

I also remember the Liberator bomber remaining at Coolham for a while and also that Michael Tanner-Garbarz made me a model of a V1 flying bomb, a V2 rocket and some photograph frames.

The frames were made from old aircraft parts and the V1 and V2's from spent canon shells. The models also doubled as lighters.

❑❑❑❑

Exploring the area with his brother, also on bicycles, is recalled by Richard Dayton-Lewis.

My brother and I lived at West Chiltington Common near Pulborough during the 2nd World War.

Coolham was just over a mile away to the north east, a simple cycle ride for two 12 year olds. And this was the way we passed our holidays, exploring on our bikes. There was no traffic to hinder us and we went for miles.

We cycled to the airfield at Coolham pushing our way through the hedgerow to the tents on the other side and started to make friends.

These men spoke a repetitive phrased English, however, youngsters are good at breaking barriers and we found it all very exciting.

The aircraft that comes first to my mind now is the P-38 Lightning with its twin boom and rudders.

The Spitfires and American Mustangs though were the main part of the scene, their merits hotly contested according to which plane one favoured most.

We followed events closely and knew something very serious was going on, which was also exciting.

We had watched some of the dogfights of the Battle of Britain and had cheered when enemy aircraft scorched uncontrollably from the sky.

On one occasion at Coolham when we were sitting with our Polish friends before one of the tents, we watched a fighter plane taxi out and start its take off.

We watched it accelerating down the runway until it was obscured by the bunkhouse building now known as Farleys. A moment later there was a loud explosion. A large cloud of black smoke appeared behind the trees and the crackling of a myriad small explosions followed. (This was Skip Payne's fatal crash referred to in the chapter on 129 Squadron.)

The men around us disappeared but soon returned to shush us away from the field and send us all home.

They explained that the pilot had failed to clear the trees at the end of the runway and that ammunition exploding in the fire was dangerous.

❑❑❑❑

David Francis was a Sergeant wireless operator/air gunner on a weekend pass, 19/20 February 1944, when he was told a Marauder had crashed at Coolham.

I cycled over from Itchingfield to see it. A sergeant I spoke to told me that the man on the ground that was killed was in charge of a unit constructing the airfield and had he thrown himself flat as the others did when they saw the plane coming in, it would have missed him. As it was, one of the propellers hit him.

❑❑❑❑

Robin Woodman recalls going from Horsham by bus with his mother to visit the Cherrimans in Sproutes Lane, Coolham.

Like many a schoolboy I was fascinated by aircraft and could not believe my eyes when I discovered an airfield just across the road.

I could hardly wait to get down from the tea table to go and investigate.

Told not to wander too far and just to have a look I saw some airmen sitting around on the grass on the edge of the airfield.

They noticed me and beckoned me over. I clambered through the wire and sat down on the grass with them. I was immediately given a large mug of strong tea, also a very thick-cut cheese sandwich to eat.

The airmen did not seem to speak much English but were very friendly and appeared to like me being there.

I managed to eat all of the sandwich and drink the tea. Shortly after this I raced back over the road to tell of my adventure.

I can still clearly remember that great cheese sandwich and the pint mug of tea but for the life of me I can't remember any of the delicacies I was given to eat at that fine tea table indoors.

❑❑❑❑

More childhood memories came flooding back to Ron Rapley, now of Queensland, Australia, who lived in Ivy Cottage, near the airfield.

After it was constructed I remember going to bed one night with absolutely nothing on the airfield and when we woke up in the morning it was covered with aircraft. We never heard a thing yet they were all there, ready to go.

I also remember gliders coming down one night, missing the airfield and landing all over the place, also a German bomber being shot down and blowing up on impact almost behind Brook Cottage.

Another incident that was quite interesting to us kiddies was of two Mustangs taking off, one just behind the other, and one coming up over the lake and not quite making it so he let his bombs go and one of them went across the road, hit the road and skimmed off and went through the hedge and across into the field the other side.

The other bomb went into the other paddock alongside it but neither exploded, fortunately.

Shortly after this another Mustang crashed on take off, ending in the middle of the cricket field and the pilot was burned to death.

I also recall the Marauder bomber crash landing and the Liberator. I had the good fortune to talk to the Liberator's pilot whom I remember walking the whole length of the runway to convince himself that he could take off.

He rattled down the runway like a mad thing and I stood the other end. He, in fact, made it up but the two wheels that were still on the ground once the tail put up went through the hedge on the end of the runway where you come across Coolham Bridge.

There was more excitement when a German bomber dropped a string of bombs.

One bomb landed harmlessly in farmland, another one hit the airfield but didn't do any damage and a third went through Cohan's house, Old House as it was called. It went through the front of the house towards the lake, went through the crazy paving and in hitting this somehow it turned and went back underneath the house and finished up under the dining room.

Very fortunately it didn't explode. I can well recall that being removed. I saw the men dig it out and they took it across the other side of the lake to what used to be known as The Park and it was exploded.

I also remember that me and my pals used to consider one of the women in the village was a German spy as she always had her porch light on every night and every night they had to go down and tell her to put it out!

The only other amazing thing is that when everything had finished and was done with at the airfield everything disappeared almost as quickly as it had come.

❑❑❑❑

Mrs Vera Matthews' father was F/Sgt J. R. Markley of an RAF airfield construction and maintenance unit.

While he was stationed at Coolham myself, my mother and brother Ken moved from our bomb damaged bome in Kent to a cottage at Falconers Farm just off the edge of the airstrip.

I can remember my father used to bring other airmen home to tea there.

There used to be a single pole barrier across the lane on the Shipley side. This used to be lowered when a plane was due to land or take off.

Ken, who was 15, adds: "I found the airfield an exciting place. With a friend I used to spend quite some time foraging in the vicinity.

"I remember my father talking about a construction man being killed by a plane landing but I suppose the highlight of the period for us lads was the landing of the Liberator.

"Myself and friend climbed all over it and helped ourselves to a few souvenirs – long since lost (those that we didn't have to give back!)"

❑❑❑❑

Frank Kirkham was in the RAF but when on home leave had a chance to see what was happening around Coolham ALG.

My wife took over the job of looking after Andrews Hall and Shipley church and lived in the cottage at the end of the hall so she was well placed to see the coming and going of the aircraft.

One building was a pumping station where petrol was pumped into mobile petrol bowsers which would then go to dispersal and refuel the aircraft. The bulk of the petrol was in underground tanks.

The Americans dropped about 80 gliders close to the

Coolham airfield in and around the fields by the pub the Blacksmith's Arms, now the Countryman Inn.

I came home on leave the next day and my wife and I went for a walk round to Broomers Corner and could see a good few of them.

They were then snatched by Dakota aircraft and taken back to base, except for those which came to grief.

❑❑❑❑

The death of the airfield construction worker is recalled by Arthur Jonah Denman.

I was delivering bricks for Sussex and Dorking Brick Co to the site and at the time part of the RAF Regiment were there working on the layout of the airfield.

There was a plane making an emergency landing on the day I was there. An RAF man was in the path of the plane and was killed. The plane came to a halt at the bottom end of the field across the road going under telephone wires.

Myself and others went to the plane to see if the crew were safe. They all got out with a few minor injuries.

I'm 80 years old and I still remember the dead man laying in the field under a white sheet.

❑❑❑❑

Activity at the airfield gradually built up and crashes and other incidents were witnessed by Philip Betteridge.

When I first went up to the airfield with my brother there wasn't much going on. The planes taking off and landing did not carry bombs but on later visits there was much more activity and the planes were carrying bombs.

I remember a B-26 Marauder standing at the very southern end of the airfield to the south of the Shipley Road and a Mustang fall apart in the air and dive into the ground somewhere to the south of Shipley.

The noise it made breaking up was like the ripping sound thunder makes sometimes.

At the time this happened there was a Canadian pilot with a shotgun under his arm walking with others down the road. He turned to my brother and I and jokingly said, "I'm not going up in one of those things again," or words to that effect.

We wanted to go off on our bikes to find the wreckage but the air force control men advised us not to go.

On another day I was sitting on the fence watching the planes landing when one plane veered to one side when the wheels touched the ground. One wheel seemed to be jammed.

He eventually got down by tilting the plane to one side, coming on one wheel, slowing down before coming down on the other one, then skidding round in a circle.

One afternoon, on arriving at the airfield, there were vehicles and a gang of men on the south runway. We could hear planes taking off on the other runway so we rode on up the road and watched them through a gap in the hedge.

That was when we saw two planes collide on take off in the crosswind.

We were told by one of the men around that a plane had done a wheels up landing on the south runway that morning.

One funny little incident was an elderly lady on her bike who turned into the Shipley Road. One of the control men ran out to stop her. She did not stop and he chased her right across to the other side.

The men I refer to as control men used to sit at a table beside a pick-up truck at the south west corner of the airfield about 50 yards in from the road and sometimes one of them would stand on the crossroads.

On most of my visits to the airfield I used to sit on the fence just north of the crossroads just watching anything going on.

Sometimes airmen would talk to me. They were a very friendly lot.

❑❑❑❑

Amusing anecdotes are recalled by Colin Stockford who was a schoolboy travelling daily to Horsham and back. His parents lived in the bungalow One Oak next door to the garage opposite the Selsey Arms.

My mother was of a nervous disposition and had always liked to imagine that the Coolham crossroads were the hub of transport for southern England and therefore likely to be the object of Luftwaffe attention. The arrival of an airfield so near us confirmed her fears.

In the event the nearest attacks I can remember were a stick of bombs, presumably jettisoned by a bomber eager to get back to Germany quickly, of which the nearest fell some field or two away while I was in bed wondering whether I ought to get under the bed, since the explosions were getting nearer, and a number of incendiary bombs which fell somewhere near Slaughter Bridge and were promptly dealt with by Peter Powell and his firefighters.

We did see the effects of a V1 at Barns Green on the way home from school one day and also I managed to get a pass for Geography in the equivalent of the GCE because we had to take cover as a V1 cut out near the school during the exam and we were given some extra credit by the examiners as a result!

I also recall the Liberator arriving and saw it depart safely over Shipley.

But one of my most vivid memories is of sitting with the runway controller occasionally as he pointed either a green or red light at the incoming fighters on their return home over the Thakeham Road and also occasionally a red Very pistol being fired to stop them landing.

No doubt my being there was very improper but the airfield did not need protection from friendly locals!

I remember a number of Polish airmen visiting our home in an evening as an alternative to their huts on the airfield or the Selsey Arms opposite.

❑❑❑❑

Madge Satterly was in the Women's Land Army staying on a farm in the vicinity when Coolham ALG was operating at its peak.

I was one of the girls invited to the social functions put on, some by poultry farmer Philip Hales, for the airmen.

Many were Polish and unfortunately, because of the language problem, it wasn't easy to converse. Nevertheless we did manage to have a pleasant time and tried to forget the war briefly.

What I do remember was how wonderfully courageous these young men were and the good spirits they maintained, even under what must have been great duress at times.

My life was saved one night when, on returning from Brooks Green, a friend who lived in the village invited me in to have a hot drink before cycling the rest of the way back to Brook House.

Normally, I wouldn't have stayed, this night I did.

Soon after there was an almighty explosion. I found out later a doodlebug had landed in Falconers Field, alongside the lane I had to travel – a near miss indeed. Nobody was hurt, only a large crater in the field.

These doodlebugs went over the farm frequently, one could hear them coming from far away and got used to the throb of their engines and prayed when the engine cut out it would land in an open field.

Strangely, we took these events in our stride, even the happenings prior to the D-Day landings which brought work to a standstill on the farm and subsequently much disruption.

One day, John Strickland, the farmer, received from the War Office notification that on a certain date all the cattle and livestock had to be kept in to leave surrounding fields empty.

The object of this exercise was soon to be realised. It seems the fields in this area were similar to those of Normandy and I gathered that there was to be a drop overhead of, I think, 48 gliders to ascertain whether they could be landed in small fields which in those days had hedges surrounding them. Half were British and the others American. I think the British ones were Horsas. The American ones were Hadrians.

On the day in question, we had Top Brass from the Allied HQ's down to view the operation and there was great activity around the farm. In preparation, a film cameraman installed himself on the top of a hay rick, a wonderful vantage point.

I was delighted to be asked if I would like to view from the rick. I learned that the cameraman had been a Hollywood cine camera operator on civvy street and came from Texas.

At the appointed hour there was a roar of engines in the sky. We looked to the horizon to see what for me was an awe inspiring sight – Dakotas towing gliders that at a given point were all released and came silently down to land. Two landed in fields nearby, a British Horsa and an American one. I don't know if any of the pilots were killed. I heard some had bad landings and possibly there were some injuries.

The aftermath of this momentous event lasted a considerable time for the gliders had to be retrieved. This was taken over by the American army and gates were removed to make way for large transporters.

The army units (Sappers) moved in to camp in our woods and following on from this planes dropped rations by parachute.

I subsequently learned that many of the glider pilots were Texans. I was full of admiration for their skill and courage, especially when one US glider was removed by the pick up method.

It entailed having a thick rope of what I believe was made of pure silk, hooked to the nose of the glider. The rest of the rope was looped over what looked like goalposts.

The Dakota which was to do the pick up was fitted under the belly with a hook. The Dakota flew in very low. The first run in was a miss, the second time the glider came away and back to earth. The next time I held my breath and this time it worked and the plane and glider soared into the air.

This pick up method was only suitable for the US gliders which were lighter than the Horsa.

For me, it was all a never to be forgotten experience. But it's hard to believe when one drives around Coolham today that it was such a hive of activity at the time of the preparation for the D-Day landings.

❒❒❒❒

David Charman sets the seal on villagers' memories of the airfield adding what he remembers of what happened to it after the war.

As a child I well remember being taken to see the planes take off and land.

There were Spitfires, Mustangs, two Thunderbolts, two Lightnings and an Oxford trainer, also the Marauder and Liberator that crashed.

After the war the airfield was used as a training area for pilots in Tiger Moths who flew down to Coolham and practised at low level but did not land.

This stopped after one of the planes crashed near Shipley Windmill killing both on board.

19

The Commemorative Weekend

The culmination of Paul Hamlin's efforts in researching the history of Coolham ALG was the Coolham Airfield Remembered Community Commemorative Event and Veterans Reunion on Saturday and Sunday, 11 and 12 June 1994.

It was 50 years since D-Day, 6 June 1944, and for the commemorative weekend Paul drew together nearly 70 former airmen – pilots, ground crews and other support staff – from all over the world, together with their families.

Paul had laid on a social evening, a dedication ceremony at the unveiling of the memorial outside the Selsey Arms public house, flypasts by a Spitfire and P-51 Mustang and a remembrance service.

For the first time in 50 years the airfield was "live" again with vintage aircraft flying in, ground displays of airfield memorabilia, a band playing, Polish dancers performing and a local amateur dramatic group entertaining.

Glorious sunshine crowned the whole weekend and villagers, guests and those who had helped make it all happen, poured on to the site, the Community Commemorative Event being free to enter. It was a momentous occasion for this small Sussex village.

The first ripple of excitement and expectancy went through the crowd as the memorial dedication ceremony started with two minutes' silence at 12 noon on Saturday outside the Selsey Arms at Coolham crossroads.

The polished black granite memorial with gold lettering glinted as respect was paid to the fallen it commemorated.

Present were former airmen from Coolham ALG, representatives of civil and squadron associations, embassy officials and military attaches, civic representatives and members of the public.

Officiating clergy were Canon Antoni Kapuscinski, chief padre Polish Air Force Association and parish priest of the Polish Roman Catholic Church of Our Lady of Czestochowa in Nottingham, and the Reverend Daniel Legh Pope, vicar of St Mary the Virgin, Shipley.

They conducted the service in Polish and English with the Rev Legh Pope dedicating the memorial to the "gallantry" of those who fell.

Twenty-three wreaths were laid and Horsham Borough Silver Band, under conductor Colin Hogg, played for the hymns, *O God, Our Help In Ages Past,* and *Immortal, Invisible.*

A Spitfire flew over in salute, causing hearts to flutter at the sweet sound of its Merlin engine, and the third time it went into a steep climb before heading away and the Royal British Legion Standards were raised.

The Last Post and Reveille were played by David Lampard, a trumpeter of the Royal British Legion.

As the guests moved away to the nearby airfield, the band played *Sussex by the Sea.*

The wreath laying ceremony at the memorial was led by Paul Hamlin followed by Mrs Stephane Walker, Chairman of Horsham District Council and others including: representatives of the Polish Air Force Association; 306 and 315 Polish Fighter Squadrons; 129 (Mysore) RAF Squadron; 222 (Natal) Groundcrew Association; 349 (Belgium) RAF Squadron; 485 Royal New Zealand Airforce Association; Crew of B-26 Marauder Flounder Gus; Crew of B-24 Liberator Mad Madeleine; RAF Airfield Construction Branch Association; RAF Airfield Construction Officers Association; RAF Regiment Association – Sussex Branch; 49 Maintenance Unit, Faygate; Royal Air Force Association, Horsham; Canadian Veterans Association; Normandy Veterans Association, Crawley & District Branch; No. 2 Group Royal Observer Corps Association; Women's Royal Voluntary Service and 1015 Squadron Air Training Corps, Horsham.

Participants at the dedication ceremony were – Embassy officials and Military Attaches: Lt Colonel Henryk Blaszczyk, Assistant Military Attache, Embassy of the Republic of Poland, London; Wing Commander G.J. Sibbald, RNZAF, Air Attache, New Zealand Embassy; Colonel J.F. David, Canadian Defence Liaison Staff, Canadian High Commission, London.

Representatives of civil and squadron associations: S/Ldr

(Ret'd) Frank Kornicki, Polish Air Force Association; Zygmunt Jelinski (S/Ldr Ret), representative 306 Dywizjon Mysliwski "Torunski"; Bozydar Nowosielski-Slepowron (F/Lt Ret), representive 315 Dywizjon Mysliwski "Deblinski"; Arthur C. Leigh, DFC DFM, Battle of Britain Clasp (F/Lt Ret), representative 129 (Mysore) Squadron; David Cox, DFC, Battle of Britain Clasp (S/Ldr Ret), Commanding Officer 222 (Natal) Squadron, RAF Coolham; Norman Murtagh, secretary 222 (Natal) Groundcrew Association; Pierre Erkes (F/Lt Ret), nominated representative 349 (Belgium) Squadron; Derek Edwards, nominated representative 485 (New Zealand) Squadron Association; Arthur R. Hanni (Captain USAAF Ret), Richard H. Lightfine (Captain USAAF Ret), Hank Sobczynski (Lt Col USAAF Ret), representatives of 323rd Bombardment Group, 454th Bomb Squadron, 2nd Air Division, 9th Air Force, Earls Colne, Essex; Russ Journigan (Lt USAAF Ret), Frank McPhillips (Lt USAAF Ret), representatives of 491st Bombardment Group, 855th Squadron, 2nd Air Division, 8th Air Force, North Pickenham, Norfolk.

Ivor Turnnidge, President & Secretary, Royal Air Force Airfield Construction Branch Association; Air Vice Marshal J.P.R. Browne, CBE, President, Royal Air Force Airfield Construction Officers Association; Harry T. Hathaway, Honourable Secretary, The Royal Air Force Regiment Association; Tom Adams, Sussex Branch Secretary, The Royal Air Force Regiment Association; Ronald M. Cook, representative 49 Maintenance Unit, RAF Faygate; Ted Tester, President, Royal Air Forces Association, Horsham; Pearce Mercer, Secretary, Canadian Veterans Association of the United Kingdom; Douglas Kay, Secretary, Normandy Veterans Association, Crawley and District Branch No. 74; Roy Bent, Observer Commander, No 2. Group, Royal Observer Corps; Ann Pennington, Women's Royal Voluntary Service, Emergency Services Organiser, West Sussex; F/O Chris Feltham, RAFV (R and T), 1015 Squadron, Air Training Corps, Horsham.

Civic representatives: Peter Longley, OBE DL, High Sheriff, West Sussex and Mary Longley; Cliff Robinson, Chairman, West Sussex County Council; Stephane Walker, Chairman, Horsham District Council; Sir Peter Hordern, DL, MP for Horsham; Madron Seligman, MEP for West Sussex; Paul Whitehouse, QPM, Chief Constable Sussex Police; James R. Green, MBE, Chairman, Shipley Parish Council; Richard Claydon, Chairman, Thakeham Parish Council; The Earl of Lytton, Chairman, Horsham Chamber of Trade and Commerce; Bill Whitlock, Station Officer, West Sussex Fire Brigade; William Stanley Parsons, BEM (Mr Horsham).

❑❑❑❑

Visitors to the airfield on Saturday, after the dedication ceremony, were greeted by a P-51 Mustang which swept low, soared up into the sky, twisting, then swept low and did a hedge-hopping, heart-stopping run over the airfield.

Then, to the sound of recordings of wartime band leader Glen Miller and his orchestra, the VIP's and public gathered in a marquee.

This was the chance for the visiting dignitaries to have their say.

Peter Longley, High Sheriff of West Sussex, opened the commemorative event saying: "None of us would have been here today if there had been a different outcome of the events in 1944".

Referring to Paul Hamlin and thanking him "for getting us here" Mr Longley said: "We have to admire him for his historical research, which apart from anything else has put Coolham on the map".

Sir Peter Hordern, Horsham's MP, welcoming representatives from abroad and some of the same men who were stationed at Coolham in 1944, referred particularly to Wing Commander Stanislaw Skalski, from Poland, whose unit was known as Skalski's circus.

Skalski (now General Skalski) was present over the weekend and Sir Peter said: "He downed 21 German aircraft, the largest amount of any Pole. He was a hero. Without those Polish pilots I think it is true to say that we would not have won the Battle of Britain and sustained the RAF up to the D-Day invasion.

"We salute them today for all they have done for our country."

Sir Peter praised the "hard work and endurance" of Paul Hamlin and those who helped.

Sussex Chief Constable Paul Whitehouse said: "I feel exceptionally privileged and proud to have such officers as Paul Hamlin and many others here today in their off duty time to honour those who went before and let us continue to live in this country in a way we so treasure".

Polish Air Force Association representative, S/Ldr Frank Kornicki, said Poland was the first to say "no" in the face of the German onslaught.

"When others surrendered, Britain didn't. It fought and won a decisive Battle of Britain and went on fighting. The Poles helped and continued alongside the British on sea, land and air on all fronts".

He too referred to Skalski's exemplary record but added: "They were all courageous, skilfull, determined young men. The monument is a fitting tribute to the 15 airmen killed. It will remind future generations what happened here, when and why."

Praising the initiative of Paul Hamlin, supported by his committee and the people of Coolham, S/Ldr Kornicki added: "The Polish Air Force salutes you".

Air Vice Marshal J.P.R. Browne, President of the Royal Air Force Construction Officers Association, said: "We are here today to remember those to whom Coolham was home, however temporarily, while they played their part in that momentous liberation of Europe.

"This coming together of a wide variety of nations in this great cause has always impressed me.

"And Coolham, a fairly small cog, epitomised that. We have here today, representatives of those nations. Paul, we thank you, your team and the village."

Then taking centre stage himself Paul Hamlin said: "When I became interested in Coolham's long forgotten airfield, very little information was around.

"But as I started to trace airmen some extraordinary stories of endurance and heroic acts emerged. It seemed a terrible oversight that so much had happened on our doorstep yet there was nothing to indicate that those events had ever occurred. It prompted me to form a committee with the intention of redressing this.

"These events this weekend are a tribute to the airmen, both local and from far afield, who fought for this country.

"We will never forget the debt we owe to the airmen who gave their lives, 15 oak trees planted by the side of the airfield, and the memorial, will serve as a reminder."

Paul was presented with the Zloty Gold Medal as a Protector of Places of National Memory by Lt Colonel Henryk Blaszczyk, Assistant Military Attache, Embassy of the Republic of Poland, London.

Mrs Margaret Hazell, of Shipley, presented Paul with her late husband's "wings" presented to him by the Polish Air Force in recognition of his work as a flying instructor to Polish pilots, and now to Paul in recognition of his work.

❑❑❑❑

There was plenty for the VIP's, public and airmen to do and see as they spilled out of the marquee on to the airfield.

Taking pride of place and causing a stir were the 22 vintage aircraft assembled including Tiger Moths, Austers, a Miles Magister trainer and Piper Cubs.

Nearly all the ex-airmen had a flight in one and many took the controls for the first time in 50 years.

Among those enjoying themselves were Dick Lightfine, the pilot of American B-26 Marauder bomber Flounder Gus. Dick met up with his engineer, Art Hanni, for the first time in 50 years.

Hank Sobczynski, Dick's former co-pilot, had kept a slice of the wrecked bomber's fuselage with its name and the dog it was named after painted on it and brought this over for the commemorative weekend. It was much photographed and admired.

Russ Journigan, the bombadier-navigator of the B-24 Liberator bomber Mad Madeline, had a great time during his visit.

When the B-24 had to make its emergency landing schoolchildren dashed out for a closer look and Russ, kitted out in his original uniform, asked to visit the William Penn Primary School in Coolham to meet today's generation.

"The children coming out meant so much to us," Russ said, himself once a pupil of the William Penn School in California, and he was a great hit with today's youngsters.

For those strolling round the airfield there were many displays and exhibits to catch their eye and interest.

Among the most interesting was the memorabilia collected from the airfield, located by Mick Monnery and Christopher Lee. Over the previous 18 months they unearthed ammunition and various bits off the aeroplanes and Christopher said: "The airmen must have been making bullets into table lighters as several were unearthed".

His favourite item was a little cigarette lighter made out of two buttons, one Polish and one RAF. This collection may be offered to Horsham Museum.

Colour, music and dance was brought by Zywjec Polish Folk Song and Dance Group of London, and appropriate wartime type sketches and songs by Billingshurst Amateur Dramatic Society.

Well known Horsham aviation artist, Cedric de la Nougerede, who painted the Memorial Fund Commemorative Print, which raised funds, had an exhibition of his paintings.

There was also an exhibition mounted by Paul Hamlin of many wartime photographs of the airmen, their machines and activities at Coolham ALG, that rolled back the years.

After all this nostalgia there was more to come on Saturday evening, the social highlight of the weekend, when the former airmen, their wives and other invited guests attended a buffet and dance at Broadbridge Heath Leisure Centre, near Horsham, put on by Paul Hamlin.

There were many hoarse voices by the time it finished as the airmen mingled, swapped stories, reminisced and couldn't believe it was half a century since many had seen each other.

❑❑❑❑

Sunday morning at Coolham Airfield everyone was back in the marquee for the Interdenominational Service of Remembrance at 11.30am.

Officiating again were Canon Antoni Kapuscinski, chief padre Polish Air Force Association and the Reverend Daniel Legh Pope, vicar of St Mary the Virgin, Shipley.

The church choir also sang and music was by Slinfold Concert Band under conductor Malcolm Gay.

In his address the vicar paid tribute to Paul Hamlin "for the single mindedness to pursue this to the end".

Remembering the people who flew from the airfield he said: "They could have been under no illusions. The war had gone on long enough to do away with any hope that it was going to be easy or cheap in terms of life to win it.

"Those at Coolham knew the task was not going to be easy and the victory costly.

"Nothing is worth living for that isn't worth suffering for."

He praised the "courage, character and endurance of the men at Coolham" and all who took part in the Second World War.

Hymns sung were *Fight The Good Fight, Praise My Soul The King of Heaven,* and *Jerusalem.*

The prayers, blessings and address were in Polish and English and the national anthems of Poland, Great Britain, Belgium and United States of America were played.

At the start of the service standards had been laid up on the drums in front of the dais by Roy Taylor, Horsham RAFA; Sgt Neil Worth, 1015 Horsham Squadron ATC; Tracey Parker, Billingshurst Royal British Legion and Syd Gander, Cowfold Royal British Legion.

The airfield remained open for the rest of the day.

Ex Coolham airmen who attended the commemorative event (country from where they travelled in brackets, all others from UK):–

133 Wing, 84 Group, 2nd TAF: General Stanislaw Skalski, DSO, DFC, VM (Poland).

315 (Polish) Squadron: pilots Jakub Bargielowski (Australia), Bronislaw Czerwinski, Dr Bozydar Nowosielski-Slepowron, Jerzy Polak, Tadeusz Slon; ground crew, fitters Stanslaw Dzialowski, Peter Skrzypczak.

306 (Polish) Squadron: pilots Zygmunt Jelinski (USA), Zygmunt Kawnik (France), Witold Krupa, Lech Laszkiewicz, Karol Marschall, Stanislaw Rudowski-Marlor; ground crew, fitters Kazimierz Barczak, Tadeusz Jakubowski; armourer Michael Tanner-Garbarz; instrument technician Tadeusz Krzystek.

411 (Polish) Repair and Salvage Unit: fitters Stanislaw Gogolowski.

5090c (Polish) Mobile Signal Unit: radio technician Stanislaw Czajkowski.

129 (Mysore) RAF Squadron: pilots John Robert Byrne (Australia), Richard J. Conroy (Australia), G.R. Dickson, Fred Holmes (Canada), Arthur Charles Leigh DFM, DFC, Leslie Lunn, Peter Nicholson, Anthony F. Osbourne DFC, George Pyle, Neil Roggenkamp (Australia), A.B. Thomson (Canada), Mervyn Young; Squadron doctor Professor J.K. Mason.

135 Wing, 84 Group, 2nd TAF, 135 Wing's Operations Officer, W.E. Francis.

222 Squadron: pilots David G.S.R. Cox, DFC, R.H. Reid; ground crew, Ronald Clark, Joe Crawshaw, Dave Cummins, Dave Davis, Pete King, Lance Lidgett, Norman Murtagh, Frank Webster.

349 (Belgium) Squadron: pilot Pierre Erkes (Belgium).

485 (New Zealand) Squadron: ground crew, Derek Edwards, Philip Mechan, Stan Wilson.

5003, 5004, 5018, 5027, 5206 Airfield Construction Squadrons: Fred Hill, W.A. Winchester.

2722, 2829 Squadrons, RAF Regiment: Harry Matthews, J.B. Kennard.

49 Maintenance Unit, Faygate: Fred Compton, Ronald Cook, Bill Lewis.

Crew members of B-26 Marauder bomber Flounder Gus: Arthur Hanni (USA), Richard Lightfine (USA), Zdzislaw Sobczynski (USA).

Crew members of B-24 Liberator bomber Mad Madeleine: Russ Journigan (Hawaii), Frank McPhillips (USA).

Civilian member of staff contracted to work at Coolham ALG as grass cutter: Donald Mitchell.

ENSA South East Area: Alan Woolven, accordion player.

Emergency services employee, West Sussex Fire Brigade, Horsham: Ken Parfitt.

The following organisations and individuals exhibited or performed at the Coolham Airfield Remembered Community Commemorative Event and Reunion: Ahmed Ramiz – model of Coolham Airfield; Alan Brown Collection – Advance Landing Grounds of New Forest and Holmsley South Airfield, Hampshire; Billingshurst Dramatic Society; Brenzett Aeronautical Museum Trust, Kent; Cedric de la Nougerede and friends, aviation art exhibition; Choir of St Mary the Virgin, Shipley; Friends of the Second Air Division Memorial, Norwich; Friends of Viscount Stephen Piercey; Harris Old Time Amusements, Ashington; Horsham Amateur Radio Club; Horsham Veteran Fire Engine Preservation Group; members of the Guild of Sussex Craftsmen; Mick Monnery – metal detection enthusiast; Paul Hamlin – Coolham Airfield Remembered Photographic Collection; Paul Janicki Collection – model aircraft of Coolham; Rebel Air Museum, Earls Colne, Essex; Royal Observer Corps South Eastern Collection; Southdown Gliding Club; Sussex and Surrey Immediate Medical Care Scheme; Sussex Police helicopter; The Broadwood Morris Men; The Thomas Collection, historic forces display; The Spitfire Society, Southern Region; Vintage Aircraft Club of Finmere; William Penn Primary School, Coolham; Women's Royal Voluntary Service and Zywjec Polish Folk Song and Dance Group, London.

The following companies sponsored or assisted the commemorative event: Alfred Bull and Company Ltd, Guildford; Annetts, Horsham; Arun Fire Ltd, Shipley; B. Belchamber, Storrington; Beavacraft, GRP Design, Pagham; Big T (Tea) Ltd, Newtown, Powys, Wales; Bob Lomas, Rural Preservation Contractors, Coolham; British Aerospace, Dunsfold, Surrey; Burgoyne Alfold Insurance Services, Horsham; Cameron Power Products, Horsham; David Waller, Business Graphics, Slinfold; David Wright Signwriter, Horsham; Fast Art, Horsham; Futurama Signs Limited, East Molesey, Surrey; Godfreys of Horsham; Hall and Co, Southwater; Hawaiian Islands Spice Co; Hillview Garage, Billingshurst; Hillier Garden Centre, Horsham; Horsham Fencing Ltd; Horsham Press Ltd; Ibstock Brick Hudsons Ltd, Thakeham; J K Leech Oil and Gas Supplies, Coolham; J. Gumbrill (B.M. & P.C. Freeman) Monumental Mason, Horsham; Julian Wolfe, historic building renovator & consultant, Coolham; King & Barnes Ltd, Horsham; Les Searle, contractor & plant hire specialist, Horsham; Moulded Acrylic Products Ltd, Cowfold; Powells of Coolham Ltd; Ray Delves, surveyor, East Sussex; Roy Spiers Home Improvements, Billingshurst; Rusper Hire Ltd, Faygate; R.W. Dickson, Plumbing and Heating Service, Horsham; Southern Aerofoil, Billingshurst; Sun Alliance Management Services, Community Involvement Programme; Sussex Camera Centre, Horsham; Tanbridge House School, Horsham; The Streeter Group, Horsham; The Art Shop, Billingshurst; The Model Shop, Horsham; The Gentle Gallery, Horsham and Steyning; Tidey & Webb Ltd, Shipley; Vinds Menswear Retailers Ltd, Horsham and W.G. Keyte and Sons Ltd, Billingshurst.

As a result of the Coolham Airfield Memorial Fund Appeal, the following local registered charities each received a cheque for £1,250 at a presentation evening held on Friday, 28 October 1994, at Needles Youth Club, adjacent to Horsham RAFA:

> The Royal Air Force Association Nursing Home, "Sussexdown", Storrington.
>
> Angels International, Petworth.
>
> Sussex and Surrey Immediate Medical Care Scheme, Horsham.

The guest of honour at the function, G/Capt David Cox, DFC, presented the cheques.

A professionally edited 60 minute video of the Coolham

Airfield Remembered event is available from Martin Bird Productions, Saucelands Barn, Coolham, Horsham, West Sussex, RH13 8QG. Telephone 01403 741620.

A limited number of commemorative prints and souvenir brochures are available from the publisher.

❑❑❑❑

One person unable to attend the commemorative weekend but who attended a ceremony laid on for her the following year was Christine Shiner, the daughter of Flying Officer Stanley (Skip) Payne of 129 (Mysore) RAF Squadron.

Christine was only two years old when her father, then aged 24, was killed whilst taking off from Coolham ALG on 17 June 1944.

She was only traced shortly before the commemorative weekend and having returned to Canada, where she now lives, after a holiday in Britain, was unable to get back for the ceremony at the memorial.

To compensate for this, surviving members of 129 Squadron resident in the UK decided to organise another reunion and invite Christine to attend and lay a wreath.

This took place in May 1995 and was organised by Dickie Dickson and George Pyle assisted by Police Constable Paul Hamlin.

The wreath Christine laid was from 129 (Mysore) Squadron, bought by her father's former fighter pilot colleagues, who were pleased to share their memories with her.

Also present were the vicar of St Mary the Virgin, Shipley, the Rev Daniel Legh Pope, and Joseph MacCluskey, Secretary of the Horsham Branch of the RAFA.

Accompanied by her husband, Rick, Christine said: "I have many photos of my father and his flying pals. I even have one of him holding me when I was a few days old.

"As a small child growing up in London, where there are a lot of memorials, I was always perplexed there wasn't one with my father's name on it so obviously I am pleased to be able to lay a wreath on this memorial where he is named.

"I am very proud of him."

Christine also attended a reunion dinner at the Royal Oak public house, Handcross, had a flight over Coolham in a vintage Piper Cub aircraft and was given a guided tour of the old airfield site.

Coolham Airfield Memorial Fund Committee.

From Left: David Cleverley, Coolham Garage; Mick Monnery, metal detection and exhibits; Michael Tanner-Garbarz, ex-Coolham Polish airman; Graham Rowley, Ground Events Co-ordinator; Paul Hamlin, Project Director; Major Nigel Martin-Bird; Alastair Martin-Bird, owner of airfield site; Barry Barnes, Treasurer; Captain Richard Sherwin, Aviation Event Co-ordinator; Bob Phillips, Shipley Parish Council representative; Ron Hill, Liaison Officer, Horsham RAFA.

Photo: J. Sampson

Air Vice Marshal J.P.R Browne, CBE, President, Royal Air Force Construction Branch Officers' Association, salutes the Coolham Airfield Memorial after laying a wreath.
Photo: West Sussex Gazette

Canon Antoni Kapuscinski, Chief Padre Polish Air Force Association (left), Police Constable Paul Hamlin (centre) and the Reverend Daniel Legh Pope, Vicar of St Mary the Virgin, Shipley, West Sussex.
Photo: J. Christian

General Stanislaw Skalski, DSO, DFC, (left), Police Constable Paul Hamlin (centre) and Lieutenant Colonel Henryk Blaszczyk, Assistant Military Attache, Embassy of the Republic of Poland, London. *Photo: J. Christian*

On behalf of Poland's Council for the Protection of Monuments to Struggle and Martyrdom, Lieutenant Colonel Henryk Blaszczyk presents Police Constable Paul Hamlin with the Zloty Gold Medal as a Protector of Places of National Memory. *Photo: J. Sampson*

Coolham veterans and guests. *Photo: J. Sampson*

The Flounder Gus emblem from the original fuselage with (from left) Dick Lightfine, Police Constable Paul Hamlin and Arthur Hanni. *Photo: J. Sampson*

Zywjec Polish Folk Song and Dance Group, London. *Photo: J. Sampson*

Russ Journigan, Bombardier-Navigator of B-24 Liberator, Mad Madeline, present at the memorial dedication ceremony wearing his original uniform. *Photo: J. Sampson*

Crowds view the vintage aircraft. *Photo: P. Hamlin*

The final salute – P-51D Mustang, N51RR, flown by Norman Lees, performs a low pass. *Photo: S. Woroniecki*

Christine Shiner, daughter of Flying Officer Stanley Payne, of 129 (Mysore) Squadron, lays a wreath on the Coolham Airfield Memorial in the presence of her late father's fighter pilot colleagues.
Photo: West Sussex County Times

Photo: Z. Jelinski

20

Tribute to the Commemorative Weekend

In July 1994 this letter was sent to the West Sussex County Times newspaper by Squadron Leader Zygmunt Jelinski, a Flight Lieutenant in 306 (Polish) Fighter Squadron at Coolham.

Upon my return home from Europe, I would like to pay a tribute to the Horsham resident, Paul Hamlin, for his outstanding effort in organising the 50th Anniversary Coolham Event, and for erecting a Memorial to the airmen who lost their lives while flying from the Coolham Advanced Landing Ground prior and during the invasion of the continent on 6 June 1944.

Over two years ago, he conceived an idea to do something that would remind the local population that Horsham residents also played an important part in World War II and that on a local meadow an airfield was built which was used to launch fighter aircraft in support of the Invasion of Europe.

Paul Hamlin's dream grew into a tremendous project, which he tackled with determination and with singularity of purpose. He knew that among the units stationed in Coolham at the time of the Invasion was a Polish Fighter Wing consisting of two Polish and one RAF squadron.

In an effort to find the surviving airmen he approached several Veteran associations at home and abroad. He researched many of the declassified government documents from that period. He personally crossed continents to interview surviving pilots and members of the ground personnel. He organised The Coolham Airfield Remembered Committee and infected its members with his own enthusiasm. They in turn launched a fund-raising drive for the Memorial and many of the Horsham residents donated their time and their talents in order to make the project a success.

And a great success it was! When, as a representative of the 306 Squadron, I was invited to lay a wreath at the base of the memorial, and at that moment a lone Spitfire flew over our heads, it was as though the 50 years dissolved into nothing and again I became the fighter pilot ready for action; but even that moment was surpassed by the daring manouevres of the P-51 Mustang aircraft which graced the sky over our heads.

The field itself became alive again, with pilots from the surrounding clubs landing their small planes and offering rides for the guests. It was as though the whole town of Horsham had turned out to celebrate; and apart from a number of exhibitions which, among others, included a stand for the Spitfire Club, the gallery of Cedric de la Nougerede's magnificent paintings, Paul Hamlin's own Coolham Airfield Remembered photographic exhibition, there was even a small fairground for the children.

Yes, Paul Hamlin made his dream come true, but it was more than that – he brought the whole community of Horsham together, which demonstrated not only their talents and underlined their achievements in a number of commercial areas, but also it emphasised their civic pride by taking part in such a magnificent project.

I am sure that Paul must have been very proud that the people of Horsham have supported him so wholeheartedly, not only by volunteering their services, but also by contributing generously to the Memorial Fund.

Paul Hamlin proved to everybody, including himself, that he was a man of vision and of unwavering determination. As a pilot in the 306 Polish Fighter Squadron, I am very grateful to him for reminding us that, at that time, the whole of England was the fighting ground, and that the glory belonged not to us, who were in uniform, but to all the British people, who also were in the front lines and who sacrificed and sufferred hundreds of thousands of civilian casualties.

I am sure that Horsham is proud of its native son, and as for myself, and the few of us that are still left, our hats are off to him.

Zygmunt Jelinski, S/Ldr (Ret'd),
California, USA.

21

Reference Diary of Events

This record of events relating to the earliest days of Coolham ALG, its operational life and sorties, was compiled by Paul Hamlin from official documents and sources. It continues after the commemorative weekend during which he put the finishing touches to the project to permanently sign the site and have the memorial listed.

Sunday, 26 September 1943: A Voluntary Detachment of personnel from the airfield attended a Battle of Britain Remembrance Service in Billingshurst Parish Church. Afterwards a parade was held under the command of F/O Ludgate. The march past consisted of the RAF, Royal Observer Corps, Home Guard, Air Training Corps, County Police and Civil Defence. S/Ldr J.K. Lancaster took the salute and complimented W/O Chapman on the smart appearance and turnout of personnel from Coolham.

21 November 1943: Two Coolham based airmen were arrested by Police Constable Porter and jointly charged with stealing two Light Sussex cockerels belonging to George Bryant of Ivydene, Shipley. At Horsham Petty Sessions on **21 January 1944**, one of them admitted the charge stating that he and his fellow airman were broke and thought they might be able to get a chicken to exchange for cigarettes. "As luck would have it, we were caught!" Presiding magistrate Mrs G. Courage commented: "There is far too much of this chicken stealing going on in the neighbourhood." The airman was fined £2.0s.0d with 12s 8d costs. The other airman did not appear and a warrant was issued for his arrest.

15 February 1944: At 11.00hrs the first landing and fatality occurred when B-26 Marauder Bomber, Flounder Gus, Serial No. 41-34913 of 323rd Bombardment Group, 454th Bomb Squadron, 2nd Air Division, 9th Air Force, Earls Colne, Essex, force landed. The aircraft hit Sgt Charles Gordon, a member of 5018 Airfield Construction Squadron. The young American air crew consisted of Lt Richard Hatherly Lightfine, aged 21, pilot; Lt Arthur R. Hanni, Bombardier-navigator; Staff Sgt Fay C. Kruger, engineer-gunner; Staff Sgt Rue Farnum, radio operator-gunner, and Staff Sgt Lloyd A. Raymond, tail gunner. The aircrew miraculously survived the forced landing after a wing collided with an oak tree. The fuselage broke in half, soaring over Thakeham Road and landing in Bridge Hill Farm. Staff Sgt Rue Farnum was knocked unconscious receiving a broken nose, while Staff Sgt Raymond sustained minor cuts to his face. They were taken by ambulance to the Canadian Base Hospital at Roffey, Horsham, where both injured airmen were detained for 24 hour observation. The remaining aircrew were released, taken to RAF Dunsfold and flown to their base.

26 March: No. 8 Section, 422 Aviation Fuel and Ammunition Park transferred to Coolham ALG under F/O Taylor and commanded by S/Ldr J.B. Batchelor. The Unit set up fuel supplies, pyrotechnic, ammunition and bomb dumps, together with oxygen equipment for aircrew use.

1 April: No. 133 Airfield, 84 Group, 2nd TAF, consisting of Polish Air Force Fighter Squadrons 306 Torunski and 315 Deblinski, transferred to Coolham ALG from RAF Heston, commanded by S/Ldr Stanislaw Lapka and S/Ldr Eugeniusz Horbaczewski, DSO, DFC. W/Cdr Stanislaw Skalski, DSO, DFC, was present. G/Capt Tadeusz Nowierski, DFC, was the Commanding Officer. The Chaplain was S/Ldr Conrad Stolarek. The following Units transferred that day:

411 (Polish) Repair and Salvage Unit under the command of S/Ldr Dabrowski tasked with the repair, service and salvage of aircraft within 11 Group. They were assisted by personnel from 49 Maintenance Unit, RAF Faygate, near Horsham, under the command of W/Cdr L.F. Chambers.

5090 (Polish) Mobile Signals Unit, 20 personnel under F/Lt Swan.

2722 (Anti-Aircraft) Squadron RAF Regiment under F/Lt D.G. King, moved to the airfield where the following gun site defences were made operational: four 40mm Bofors, eight 20mm Hispano cannons, and twelve twin 303 Browning machine guns.

2 April: British Double Summertime introduced.

3 April: 129 (Mysore) RAF Squadron, a multi-national unit commanded by S/Ldr Charlton "Wag" Haw, DFM, DFC, Order of Lenin, arrived at Coolham ALG from RAF Llanbedr, Wales. That same day a Servicing Flight consisting of Polish airmen en route from RAF Llanbedr to Coolham were involved in a major road traffic accident three miles S/E of Dolgallau, Gwynedd, Wales, which resulted in four of its 3-ton lorries overturning. Aircraftman P-794879 Irenrusz Swiatczak, aged 26, died in the accident. In addition, six personnel were seriously injured and nine slightly injured.

Within days of arriving, the three fighter squadrons relinquished their Vickers-Supermarine Spitfire IX's and began familiarisation flights on the North American P-51 Mustang. Throughout April 1944 the weather remained inclement. This, without doubt, hampered training. Occasionally, RAF Ford was used owing to it having a concrete runway. However, on **4 April**, F/Sgt Gozdzielewski of 315 Squadron, was involved in an accident at Ford while taxiing his P-51 Mustang, the propeller receiving superficial damage. Two days later, on **6 April**, at 16.20hrs, F/Sgt Ciundziewicki of 315 Squadron, had an accident at Ford. Taxiing his P-51 Mustang after landing, the aircraft left the runway causing damage to the port undercarriage. At the time the pilot was making an effort to clear the runway due to another aircraft landing rapidly behind. The subsequent investigation resulted in the pilot receiving the following endorsement in his log book: "Landing 6 April 1944 – Carelessness."

12 April: First air raid warning was given at Coolham airfield.

13 April: F/Sgt Slon of 315 Squadron was involved in an accident at Coolham. While taxiing after landing his P-51 Mustang damaged its propeller when it collided with an ambulance. No injuries were sustained.

15 April: P/O Laszkiewicz of 306 Squadron flew from Coolham to RAF Rednal on a cross-country training flight in a Spitfire Vb. While taking off the tailwheel became detached. On landing at RAF Rednal the aircraft received slight damage to its rear fuselage and mainplane tail flap.

16 April: Arrival of 16 ALG (Signals) for attachment, unit comprised of 28 personnel under F/O Martin and F/O Woolgar.

17 April: 2829 (Field) Squadron RAF Regiment transferred to Coolham under F/O Hutton. The squadron's task was security of the airfield and all nearby roads together with guarding crashed aircraft in the district. S/Ldr A. Dickinson remained in command of both 2722 and 2829 Squadrons RAF Regiments.

18 April: At 15.15hrs F/O Baker of 1 Detached Delivery Flight, 11 Group, was injured when he crashed on take off while collecting a Spitfire Vc Serial No.EE643 from Coolham. An investigation revealed that the pilot, who had an artificial leg, took off with the propellor in coarse pitch and was unable to maintain height, eventually stalling and crashing into a wooded area at Hoe's Farm, Coolham. RAF Form 1180 (Accident Record Card) states that the Spitfire sustained Category "E" Damage (written off), the pilot receiving an endorsement in his logbook, "Landing 18 April 1944 – Carelessness".

19 April: During an air raid warning at Coolham airfield, an Me410 was shot down by a night-fighter at Cooks Farm, Nuthurst, near Horsham.

Between **19 and 24 April** all three squadrons participated in "Exercise Lambourne" providing low level air support to Army Exercises.

21 April: 422 Aviation Fuel and Ammunition Park transferred to Slindon Park, West Sussex.

25 April: F/Sgt Tomczak of 306 Squadron was injured while taking off in an Auster at 15.00hrs when it swung on take off colliding with tree stumps and overturned. An examination of the aircraft revealed that the undercarriage axle had been slightly bent and this, together with a cross-wind, contributed to the accident. However, the pilot still had his log book endorsed in red ink – "Carelessness".

26 April: Coolham Airfield became fully operational. The first Ranger Mission was made by four P-51 Mustangs of 129 (Mysore) Squadron lead by F/Lt Leigh, DFM, DFC, at 14.20hrs attacking five airfields in the Albert–Cambrai area, France. Accurate and heavy flak was experienced during the initial stages of the mission which was curtailed after three aircraft were hit resulting in the death of P/O Nelson.

29 April: During Ramrod 806 a Wing sweep of Le Touquet–Cambrai–Coulommiers area resulted in S/Ldr Horbaczewski being slightly wounded, the canopy of his P-51 Mustang damaged during heavy flak.

30 April: 133 Airfield Operations Record Book (RAF Form 540) states: "The month was spent on the very strenuous work of receiving and refitting Mustang aircraft involving all technical sections."

3 May: Polish National Day. A holy mass was held for personnel not on duty.

4 May: All three squadrons participated in "Operation Fabius" providing Fighter Umbrella cover to troops training for the invasion.

12 May: F/O Ruchwaldy of 129 (Mysore) Squadron experienced total brake failure while taxiing his P-51 for take off at 12.27hrs to participate in Ramrod 875. That same day the Chief Orthodox Chaplain to the Polish Forces, Reverend Sauva, visited the airfield.

15 May: 133 Airfield was redesignated 133 Wing, 84 Group 2nd TAF.

18 May: At 17.40hrs while participating in a Ranger Patrol 12 miles NE of Nevers, an He111 bomber was destroyed in combat by F/Lt Potocki, F/Sgt Pomietlarz, F/O Beyer and F/O Tronczynski of 306 Squadron.

20 May: Four of 133 Wing's P-51 Mustangs were damaged in flying accidents. At 12.50hrs while returning from Ramrod 898, F/O Pietrzak of 306 Squadron forgot to lower his undercarriage when landing causing considerable damage to the S/W–N/E runway. This was repaired in just over four hours by personnel from 5004 Airfield Construction Squadron. Operations continued during the day using the longer E/W runway. However, a strong 90 degree crosswind made flying conditions hazardous. At 15.51hrs during Ramrod 904 W/O Roggenkamp of 129 (Mysore) Squadron crashed on take off when his aircraft swung to port. At the same time F/O Wedzik of 306 Squadron completed the same manoeuvre damaging another fighter in dispersal. All three pilots were uninjured but a ground crew member sustained slight injuries as a result of W/O Roggenkamp's fighter careering into a latrine he was using!

25 May: 19.15hrs while participating in a Ranger mission in the Bourges area, two Arado Ar 96 German training aircraft were destroyed in combat by F/Lt Marciniak, F/O Polak, F/O Kirste and F/Sgt Jankowski of 315 Squadron.

29 May: Polish C-in-C General Sosnkowski and AVM General Mateusz Izycki visited Coolham ALG and decorated Wing Commander Skalski with the Polish Order of Virtuti Militari 4th Class. At 15.00hrs on the same day, AOC 2nd TAF Air Marshal Sir Arthur Coningham, KCB, DSO, MC, DFC, AFC, arrived at the airfield in an Avro Anson. All personnel assembled in the centre of the airfield by the flying control caravan and were instructed to sit around him. The Air Marshal climbed on a jeep and gave a speech regarding the aim of the forthcoming operations: the destruction of the German army. He emphasised the following points:

(i) Thoroughness in destruction of enemy targets.
(ii) Effective ground support.
(iii) Effective aircraft dispersal against enemy attack.

1 June: Children from the local school visited the airfield. F/O Tomszyk, Junior Intelligence Officer, in charge of reception.

2 June: At 17.30hrs a tent used by 129 (Mysore) Squadron at their dispersal was burnt out. By 20.00hrs the same evening a second fire was discovered in S/Ldr Haw's admin office but was quickly extinguished.

3 June: The painting of recognition stripes to aircraft was ordered. That same day during Ramrod 964, W/O Tamowicz of 315 Squadron survived when the cockpit canopy of his P-51 Mustang blew off while dive-bombing, damaging the aircraft's mainplane, stabilizer flap, elevator and tail fin.

6 June, D-Day: 05.20hrs, a printed leaflet with General Eisenhower's Order was distributed, however, for the majority of that day, 133 Wing was not operational until allocated Ramrod 976. At 20.20hrs W/Cdr Skalski led all three Coolham based squadrons, 36 aircraft in total, as close escort to Airborne Forces landing in the Eastern Beachhead, the weather being no cloud, slight haze, visibility 15/20 miles. At 21.15hrs during the same mission F/Lt Hancock, DFC and W/O Rigby of 129 (Mysore) Squadron destroyed an Fw190 fighter 15 miles S/W of Caen.

The personal combat report of F/Lt Hancock states as follows: "The squadron were escorting Stirlings towing gliders. Blue section, which I was leading, left the squadron as arranged. We flew on a S/W course when I sighted one E/A (enemy aircraft) 6,000 feet below at 10 o'clock. I at once went down, but overshot, pulling up to port and above. Blue 2 closed and fired, I observed strikes on the wings. E/A then broke sharply to starboard, Blue 2 overshooting, E/A then flew east. I at once went down again giving two 2-second bursts, and he blew up in the air and I flew through the pieces. This was observed by all Blue section. I claim this E/A as destroyed."

7 June: Coolham ALG undoubtedly endured its most hectic and momentous day when 133 Wing flew 4 Ramrod missions commencing at 06.00hrs and finishing at 21.37hrs resulting in the destruction of 16 enemy aircraft, 3 possibles and 2 damaged. At about 07.00hrs while participating in Ramrod 980, Part I, between Evreux and Bernay, F/O Twomey of 129 (Mysore) Squadron destroyed an Fw190 and W/O Roggenkamp of the same squadron was shot down during combat and posted missing. At the same time, while participating in Ramrod 980, Part I (dive bombing) 3 miles W of Argentan, F/Lt Sworniowski of 315 Squadron destroyed an Fw190 in combat.

Between 10.10hrs and 10.20hrs, while participating in Ramrod 980 Part II, four Bf109 fighters were destroyed in combat, 15 miles SE of Rouen, by F/Lt Marciniak, F/S Berka, F/O Kirste and F/Sgt Idrian of 306 Squadron. At 10.35hrs in the Pont Audemer area, S/Ldr Lapka, F/Lt Sologub, F/Lts Budewicz, Sierkierski and F/O Beyer, each destroyed a Bf109F while in combat with 20 plus enemy fighters. F/O Laszkiewicz did not return from the mission and was posted missing. S/Ldr Lapka, the Commanding Officer of 306 Squadron, was shot down during combat but made a safe parachute descent and was fortunate in being picked up by members of the French Resistance who looked after him, eventually arranging his repatriation.

During the same mission two P-51 Mustangs of 306 Squadron were seriously damaged but were able to return to Coolham. While engaged in a low level strafing attack, a P-51 flown by F/Sgt Letki of 306 Squadron, struck a tree damaging the wing's leading edge. RAF Form 1180 states that the pilot was using natural cover against flak. No disciplinary action was taken as it was considered an "operational hazard". While dive bombing and at low level a P-51 flown by F/Lt Budrewicz sustained damage to the leading edge of its mainplane. RAF Form 1180 states that the Squadron C/O could not ascertain whether damage was caused by enemy flak or from pilot's own bomb splinters! As a result no disciplinary action was taken.

At about 18.00hrs, while participating in Ramrod 980 Part III in the Argentan–Caen area, F/Lt Potocki and F/O Pietrzak of 306 Squadron were involved in combat on two occasions during the same mission. At 18.05hrs F/Lt Potocki and F/O Pietrzak each destroyed a Bf109 in the first engagement and again, 10 miles south of Caen when F/Sgts Rudowski and Nowoczyn of the same squadron also claimed a Bf109 in combat.

8 June: F/Lt Janusz Marciniak was posted from 315 Squadron to 306 and appointed the rank of Acting S/Ldr. While participating in Ramrod 982, which commenced at 05.55hrs, a P-51 Mustang, flown by F/Lt Budrewicz of 315 Squadron, made a forced landing in the beachhead area and was posted missing. At 16.00hrs that afternoon, participating in Ramrod 982 (dive-bombing), a P-51 flown by F/Lt Cwynar of 315 Squadron, was hit by flak, force landing in Allied territory beside the Caen–Falais road and was briefly posted as missing.

10 June: F/Lt Tomanek of 306 Squadron failed to return after participating in Ramrod 986. In addition, two pilots of 129 (Mysore) Squadron, flying P-51 Mustangs, failed to return during a Ranger Mission strafing German tanks. P/O Pyle force landed in enemy territory west of Vassy and W/O Thomas, an Australian, in the middle of a German minefield, west of Bayeux.

The latter's story was recorded by a war correspondent on board HMS Hilary while returning to England the day after the crash. "As I was flying across a small French village, light ack-ack fire came up and my aircraft was badly holed. I was about 25 miles inland and immediately turned for the coast. After a few minutes flying my engine seized and started to pour out smoke. I looked for a likely place and brought her down. She went through a hedge and over two fields, stopping in the middle of a third.

"I was not hurt and clambered out of the cockpit. On the road two British soldiers were yelling at me to stay where I was. I waited for a while then a British Engineer Officer approached and guided me out of the field which was sown with mines. How I missed them I cannot think"

Prior to returning to England it was reported that W/O Russell Thomas purchased French perfume in the town of Courseulles-sur-Mer to send home to his wife in Melbourne, Australia. Unfortunately the gift did not arrive.

12 June: At 12.40hrs while participating in an Armed Recce mission eight miles SE of Sees, S/Ldr Horbaczewski, F/O Kirste and F/Sgt Bargielowski of 315 Squadron destroyed four Fw190s without loss.

13 June: At 19.45hrs while participating in Ramrod 995, five miles NE of Flers, F/Lt Stefankiewicz destroyed an Fw190 in combat.

14 June: F/Lt Byrne of 129 (Mysore) Squadron did not return from Ramrod 998 and was posted missing. At 12.00hrs, while taxiing, one of the two 500lb bombs being carried by S/Ldr Haw's P-51 Mustang fell from its wing mount, damaging the mainplane tail flap.

15 June: AVM L.O. Brown, AOC No. 84 Group 2nd TAF, visited Coolham and decorated W/O Malinowski of 411 Repair and Salvage Unit.

16 June: F/Lt Zbrozek, Public Relations Officer, and F/O Bielec, photo reporter, were posted to 133 Wing.

17 June: At 22.00hrs while participating in Ramrod 1012 Part IV, two Fw190s were destroyed in combat 10 miles SW of Nogent by F/Lt Potocki, F/Lt Jelinski, F/O Beyer and F/Sgt Pomietlarz of 306 Squadron. During the night of 17/18 June 2722 (AA) RAF Regiment used their guns for the first time at the airfield, on three occasions firing at V1 flying bombs. The following ammunition was fired: Bofors cannon – 46 rounds, Hispano cannon– 6 rounds, Browning – 60 rounds. No hits were claimed or casualties reported. Soon after the engagement S/Ldr Dickinson, Commanding Officer of 2722 (Anti-Aircraft) RAF Regiment immediately issued Orders for Engagement of Pilotless Enemy Aircraft:

(i) Bofors and Hispano cannons will only be used.

(ii) Fire will NOT be opened on direct approaches, only on crossovers or receders.

(iii) All fire be directed OUTWARDS from airfield perimeter and dispersals.

(iv) Danger area from exploding pilotless enemy aircraft will be taken to be a radius of 500 yards, targets NOT to be engaged within a lesser distance.

(v) Minimum gun crews will go into action, remaining personnel to take cover in slit trenches.

(vi) Squadron HQ to be notified immediately a pilotless enemy aircraft is sighted, or suspected."

18 June: At 23.00hrs No. 11 Group Signal message was received stating "Guns deployed for defence of airfields will in no circumstances be allowed to engage enemy aircraft". Squadron records state that as a result of the order "spirits sank to zero". However, on the night of 22 June, the instructions were rescinded and personnel were "on top of the world".

22 June: One of the most heroic acts of aerial bravery of World War II occurred during a tank strafing mission under 2nd TAF Ops Order 188, when S/Ldr Horbaczewski rescued fellow pilot W/O Tamowiecz from his crashed aircraft in a marsh near Cherbourg. Both returned to Coolham ALG in the Commanding Officer's single seater P-51 Mustang, the injured airman sitting on S/Ldr Horbaczewski's lap. During the same mission a P-51 flown by F/Lt Leigh of 129 (Mysore) Squadron was hit by flak and made a successful landing at an American Emergency Landing Ground.

At 16.45hrs, while taxiing to take part in an Armed Recce mission, a 500lb bomb being carried by a P-51 Mustang flown by F/Lt Strachan of 129 (Mysore) Squadron, fell from its wing mount causing damage to the aircraft's tail flap. An investigation revealed the bomb rack to be faulty and that a bumpy perimeter track had been a contributing factor. That same day 133 Wing transferred from Coolham ALG to RAF Holmsley South in the New Forest, Hampshire.

During May and June 133 Wing flew as a combined unit on 41 specified Ramrod missions, five of these being withdrawal cover to B-17 Fortresses and B-24 Liberators returning from daylight raids into occupied Europe. These sorties averaged four and half hours' flying time and entailed encroaching deep into enemy territory. Its squadrons also flew numerous armed reconnaisance Ranger and Rodeo missions over northern France bombing and strafing enemy targets of opportunity including railways, airfields, enemy troop transport concentrations and shipping.

While stationed at Coolham, 133 Wing lost 21 P-51 Mustangs in action and three in non-operational or familiarization flying. During the 57 days its pilots were operational they destroyed 28 enemy aircraft in aerial combat, consisting of the following:

306 Squadron:	11 Messerschmitt Bf109's
	2 Focke-Wulf Fw190's
	1 Heinkel He111.
315 Squadron:	10 Focke-Wulf Fw190's
	2 Arado Ar 96
129 (Mysore) Squadron:	2 Focke-Wulf Fw190's

It is believed that approximately 700 personnel in total served at Coolham. Fifteen died in the course of their duties, eight Polish, six British and one Canadian.

23 June: At 08.05hrs No.1 Bofor's Detachment commanded by Sgt Rhodes of 2722 (Anti Aircraft) RAF Regiment fired at, and hit, a V1 flying bomb which momentarily showed itself through a gap in thick cloud.

Although the V1 was not seen but heard to explode, a claim for its destruction was made.

30 June: 135 Wing of 84 Group transferred to Coolham from Selsey ALG, East Sussex, commanded by G/Capt Johnnie Walker, DSO, DFC and W/Cdr P J Simpson, DSO, DFC. The Wing was equipped with Spitfire LF IXe's and consisted of:

222 (Natal) Squadron commanded by
S/Ldr D.G.S.R. Cox, DFC and Bar, C de G

349 (Belgium) Squadron commanded by
S/Ldr I.G du Monceau de Bergendal, DFC and Bar, C de G

485 (New Zealand) Squadron commanded by
S/Ldr J.B. Niven, DFC.

While at Coolham a member of 485 (New Zealand) Squadron, F/O Johnnie Houlton, DFC, arranged to take a 48 hour break with his future wife, Vicki. This was curtailed due to him becoming violently ill after consuming suspected contaminated beer. An RAF medical officer who examined F/O Houlton was of the opinion that he required immediate medical treatment. The pilot was taken to the Canadian Base Hospital at Roffey, Horsham, where he spent five very comfortable days prior to returning to flying duties with his squadron which had by then vacated the airfield.

1 July: 135 Wing flew as a combined unit on Ramrod 1052 providing support and withdrawal cover to Lancasters and Halifaxes bombing No-ball targets (V1 flying bomb sites) in the Pas de Calais area.

2 July: 349 Squadron flew a low cover Ranger patrol in the Normandy area. That same day, 2722 (Anti-Aircraft) Squadron RAF Regiment transferred to Folkestone, Kent.

3 July: 349 Squadron completed a Ranger mission before participating in Ramrod 1056 with 135 Wing, providing close escort to 48 B-25 Mitchells bombing a target in the Argentan area.

4 July: 135 Wing transferred from Coolham to Funtington ALG, West Sussex. During their five day stay at Coolham the Wing did not sustain any fatalities or loss of aircraft. That same day 2829 (Field) Squadron RAF Regiment transferred to RAF Aston Down, Wiltshire.

15 July: 411 (Polish) Repair and Salvage Unit transferred to Appledram ALG, near Chichester, West Sussex. No other squadrons operated from Coolham ALG from that date although the runways and buildings were fully maintained, the grass being cut by a local civilian contrator.

19 January, 1945: The last documented landing at Coolham ALG using the Sommerfeld metal strip runways occurred during the afternoon when B-24J Liberator bomber Mad Madeleine, serial no. 44-40164 of 491st Bombardment Group, 855th Squadron, 2nd Air Division, 8th Air Force, based at North Pickenham, Norfolk, made an emergency landing. The American aircrew consisted of Lt Frank Goodrich, senior instructor pilot; Lt Norman E. Flaherty, 1st pilot; Lt Frank C. McPhillips, co-pilot; Lt Russell P. Journigan, bombardier/navigator; Tech Sgt Frank T. Eaton, flight engineer; Sgt Robert E. Lease, radio operator; Sgt Ernest Irwin, nose gunner; Sgt Frank E. Eckert, left waist gunner; Sgt Billy M. Ellington, right waist gunner, and Sgt Robert W. Mathews, tail gunner. At the time of the landing there was only one RAF Corporal present at the ALG and no rescue services.

26 January: After being repaired and relieved of all armaments in order to lighten it, the B-24J Liberator bomber finally took off during the early morning with a crew of three, completing a short ferry flight to RAF Dunsfold before being fully re-equipped and returning to its base. It has been said that when the bomber took off along the main E/W runway towards Thakeham Road, Coolham, the undercarriage made contact with the boundary hedge.

September 1945: Reinstatement of Coolham ALG to its previous agricultural status was commenced by 29 personnel from 5551 AC Flight, 5027 Airfield Construction Squadron. During **October 1945,** 22 personnel from 5003 Airfield Construction Squadron jointly assisted in the task which was finally completed on **11 November 1945.** Although the RAF had removed the Sommerfeld metal strip tracking and other moveable structures, photographic evidence has since revealed that on 4 April 1946, numerous large containers were present at the western end of the main runway. The first field to be returned to full agricultural use in the Spring of 1946 was at the eastern end of the main runway behind Knights Farm.

9 September 1946: The last official and undocumented use of Coolham ALG by an RAF aircraft occurred when a North American Harvard Trainer, serial no. KF426, landed, one of its two pilots being W/O Henrik Kowalski of 306 (Polish) Squadron based at RAF Coltishall, Norfolk. His colleague, a young Polish fighter pilot who had served operationally at Coolham, was about to return to live in Poland and desperately wanted to return to Horsham and complete "unfinished business". Early that morning during an authorised cross country navigation exercise, the Harvard landed on the grass and was left unattended for a number of hours while both airmen made their way into Horsham. After W/O Kowalski had been shopping and his colleague had successfully completed his "task" both returned to their aircraft which was still intact and flew it back to RAF Coltishall.

April 1994, 15 sapling oak trees were donated to Shipley Parish Council by Horsham District Council as part of its Parish Initiative Tree Planting Scheme and were planted alongside the Public Bridleway. Each tree bears a plaque in memory of a deceased Coolham airman.

11 June 1994: The Coolham Airfield Memorial was unveiled. It lists the squadrons who served at the ALG and the 15 airmen who died while stationed there. It is sited in the front garden of the Selsey Arms public house at Coolham Crossroads. The Imperial War Museum, London, has registered the Coolham Airfield Memorial in its National Inventory.

March 1995: The Coolham Airfield Memorial was listed with the South East England Tourist Board and is signed accordingly.

11 June 1995: A sign was officially unveiled at the airfield's northern entrance beside the B2139 Thakeham

Road by W/Cdr David Cox, DFC, ex-Commanding Officer of 222 (Natal) Fighter Squadron and Cliff Robinson, Chairman of West Sussex County Council, himself a D-Day veteran. The sign states: "Historic Site, Coolham ALG D-Day Airfield". Another sign also indicates the southern end of the ALG at Saucelands Lane.

18 April 1996: Horsham District Council presented a Civic Design Award for the Coolham Airfield Memorial to Paul Hamlin, Ron Pidgley (Calligrapher) and Julian Wolfe (Designer and Builder).

Today, the only building left at the ALG is a brick defence post situated to the east of the B2139 Thakeham Road at the southern end of the airfield site. During hostilities this building housed a pump used in conjunction with a nearby petrol installation. A recent structural examination concluded the building to be in fairly sound condition, considering its age and location. It is hoped that in the near future the Department of National Heritage will consider the defence post of Historic Interest and list it accordingly.

The perimeter track can be seen clearly from the air and sections are used as farm roads. Wartime concrete posts and fencing are also visible. Since reinstatement, partly carried out by Les Searle, Contractor and Plant Hire of Horsham, no development has taken place on the site, all land having been restored to full agricultural use.

An official West Sussex County Council Public Bridleway exists along the northern section of the old perimeter track starting at the B2139 Thakeham Road, 200 yards south of Coolham Crossroads.

Occasionally, during the summer months, a locally-owned vintage aircraft operates from a grass strip positioned over the southern section of the airfield's S/W–N/E runway.

A designated military aircraft flight corridor passes over the position of the ALG's main E/W runway. It is a common occurrence to see today's generation of fighter aircraft and helicopters passing overhead, sometimes at low level.

Shipley Parish Council has kindly accepted ownership of the Memorial and a Trust Fund Account for its future maintenance has been established. Contributions would be welcome and should be sent to the Clerk to Shipley Parish Council, Mrs Joan Nunn, Rosemary Cottage, Dragons Green, Nr Horsham, West Sussex, RH13 7JD.

A toast at the unveiling ceremony of an historic site location sign at Coolham airfield's northern entrance. From Left: Cliff Robinson, Chairman of West Sussex County Council; Mrs Margaret Hazell, Shipley resident, and Wing Commander David Cox, DFC. *Photo: West Sussex County Times*

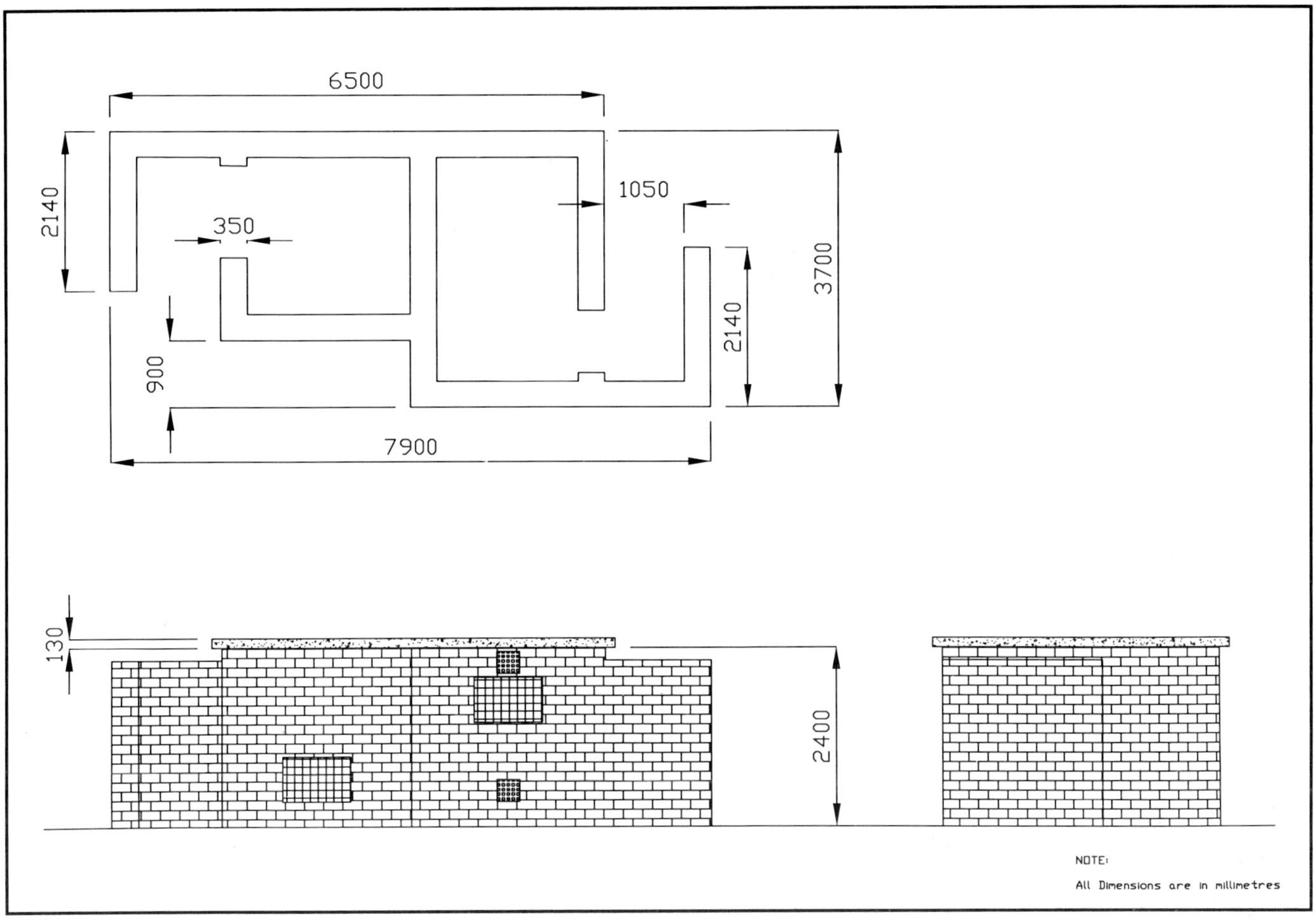

Plan of brick defence building, Coolham ALG. *Drawn by Steve A. Hough, BSc, C Eng, MICE*

Defence building, Coolham ALG. *Photo: P Hamlin*

Aerial view, 1994, site of Coolham ALG clearly showing the location of the airfield's E/W runway and perimeter track in use as a farm road, a section being a Public Bridleway.
Photo: M. McCormac

22

Squadron Personnel

The following pilots are listed in Operation Record Books (RAF Form 540) as having been stationed at Coolham ALG. They are shown in their ranks attained at the time. This is repeated at the completion of each individual's contribution. Unfortunately, records do not provide details of non flying personnel except Medical, Administrative and Chaplains.

Those marked with an asterisk (*) denote that they fought in the Battle of Britain.

133 Wing, 84 Group, Second Tactical Air Force

G/Capt Tadeusz Nowierski *
W/Cdr Stanislaw Skalski *
Chaplain S/Ldr Conrad Stolarek

POLISH AIR FORCE
306 Dywizjon Mysliwski "Torunski"
(306 Fighter Squadron)

S/Ldr S. Lapka *

F/Lt J. Bialoskorski
F/Lt R. Budrewicz
F/Lt Z. Drybanski
F/Lt B. Geca
F/Lt Z. Jelinski
F/Lt E. Krzehlik
F/Lt H. Pietrzak
F/Lt W. Potocki
F/Lt J. Siekierski
F/Lt G. Sologub
F/Lt K. Sporny
F/Lt E. Tomanek
F/Lt M. Wisiorek

F/O A. Beyer
F/O W. Figiel
F/O C. Gierycz
F/O C. Jaworowski
F/O L. Laszkiewicz
F/O K. Marschall
F/O C. Oberdak
F/O Z. Porazinski
F/O W. Szajda
F/O S. Tronczynski
F/O T. Turek
F/O M. Wedzik *
F/O A. Wegrzyn

P/O J. Bzowski
P/O Z. Kawnik
P/O J. Smigielski

W/O W. Krupa
W/O E. Loska
W/O M. Machowiak
W/O J. Rogowski

F/Sgt F. Berezowski
F/Sgt J. Czezowski
F/Sgt E. Krzehlik
F/Sgt S. Letki
F/Sgt K. Michalkiewicz
F/Sgt W. Mrozowski
F/Sgt W. Nowoczyn
F/Sgt K. Osieleniec
F/Sgt J. Pomietlarz
F/Sgt S. Rudowski
F/Sgt F. Tomczak

Medical: F/Lt J. Jarosz
Adjutant: F/O T. Bobola

POLISH AIR FORCE
315 Dywizjon Mysliwski "Deblinski"
(315 Fighter Squadron)

S/Ldr E. Horbaczewski

F/Lt M. Cwynar
F/Lt M. Gorzula
F/Lt R. Idrian
F/Lt J. Marciniak
F/Lt J. Polak
F/Lt J. Schmidt
F/Lt H. Stefankiewicz
F/Lt K. Stembrowicz
F/Lt A. Sworniowski
F/Lt S. Wandzilak
F/Lt F. Wiza

F/O T. Bobola
F/O S. Calinski
F/O A. Czerwinski

F/O T. Haczkiewicz
F/O M. Kirste
F/O P. Kliman
F/O B. Nowosielski
F/O K. Sztramko
F/O K. Wunsche *
P/O G. Swistun

W/O J. Adamiak
W/O J. Derma
W/O T. Tamowicz
W/O T. Slon

F/Sgt J. Bargielowski
F/Sgt S. Bedkowski
F/Sgt T. Berka
F/Sgt A. Ciundziewicki
F/Sgt B. Czerwinski
F/Sgt J. Gozdzielewski
F/Sgt T. Jankowski
F/Sgt A. Kepczynski
F/Sgt K. Kijak
F/Sgt Lurzynski
F/Sgt K. Siwek
F/Sgt J. Zalenski

Medical: F/Lt S. Harasymowicz
Adjutant: P/O B. Smidowicz

129 (MYSORE) RAF Squadron

S/Ldr C. Haw *

F/Lt K.C. Baker
F/Lt J.P. Bassett
F/Lt R.J. Conroy
F/Lt N.S. Green
F/Lt A.J. Hancock
F/Lt R.G. Kleimeyer
F/Lt A.C. Leigh *
F/Lt D.F. Ruchwaldy
F/Lt I.D.S. Strachan

F/O D.C. Byrne
F/O G.R. Dickson
F/O J.E. Hartley
F/O M. Humphries
F/O L.G. Lunn
F/O P.A. Nicholson
F/O A.F. Osborne
F/O D.C. Parker
F/O S. Payne
F/O A.B. Thomson
F/O M. Twomey

P/O E.W. Edwards
P/O D.H. Nelson
P/O G.F. Pyle
P/O J.R. Byrne

W/O G. Cliff
W/O A.J. Foster
W/O T. Hetherington
W/O E. Redhead
W/O W. E. Rigby
W/O N. Roggenkamp
W/O R.L. Thomas

F/Sgt W.A. Jeal
F/Sgt D.J.F. Rivett

Medical: F/Lt J.K. Mason
Adjutant: F/O W. Madge

135 WING, 84 Group, Second Tactical Air Force

G/Capt Johnnie Walker
W/Cdr P.J. Simpson *

222 (NATAL) RAF Squadron

S/Ldr D.G. S R Cox *

F/Lt R.A. Beardsley *
F/Lt D.W. Beedham
F/Lt J.B. Collins
F/Lt W.B. Gower
F/Lt L.J. Omissi

F/O W.F. Bern
F/O N.J. Buchwald
F/O A.W. Burge
F/O R.M. Davidson
P/O A.J. Lenehan
P/O R.H. Reid
W/O R.E. Drennan
W/O R.H. Oliver
W/O S.G. Walton

F/Sgt A.J. Cosgrove
F/Sgt J.D. Ford

349 (BELGIUM) RAF Squadron

S/Ldr I.G Du Monceau de Bergendal

F/Lt A. Lemaire
F/Lt G. Seydel
F/Lt A. Van de Velde

F/O P. Erkes
F/O J. Ester
F/O M. Gendebien
F/O H. Goldsmit
F/O P. Maskens
F/O A. Oger
F/O M. Renard

F/Sgt S. Blair
F/Sgt J. Bragard
F/Sgt K. Brant
F/Sgt G. Halleux
F/Sgt A. Moureau
F/Sgt D. Smerdon
F/Sgt R. Van den Bosh

Medical: F/O De Gand
Adjutant: F/O Gulson

485 (NEW ZEALAND) RAF Squadron

S/Ldr J.B. Niven

F/Lt J.N. King
F/Lt K.C. Lee
F/Lt K.J. MacDonald
F/Lt D.X. O'Halloran

F/O T S.F. Kearins
F/O M.C. Mayston
F/O D.A. Roberts
F/O F. Transom
F/O L.S.M. White
F/O J.F.P. Yeatman

P/O H.W.B. Patterson

W/O H.M. Esdaile
W/O J.J. Robinson

F/Sgt E.G. Atkins
F/Sgt A.F. Bowner

F/Sgt C. McInnes
F/Sgt W.T.H. Strahan

Adjutant: F/O A. Hutton

23

List of Fatalities

Photo: Mrs. V. Hogg

Gordon, Charles, Sgt 967357

Born in Scotland 11.12.1901, died aged 45.

At 11.00hrs on 15 February 1944 Sgt Gordon, a member of 5018 Airfield Construction Squadron, was engaged in Sommerfeld track laying duties when hit by B-26 Marauder bomber Flounder Gus which made an emergency landing at Coolham ALG while the airfield was under construction.

He is buried at Ballengeich Cemetery, Stirling, Scotland, Section M, Grave 27.

Photo: Z. Jelinski

Szajda, Wladyslaw Jan, F/O P-1740

Born Tarnopol, Poland 31.10.1919, died aged 24.

At 14.52hrs on 4 April 1944, F/O Szajda, a member of 306 (Polish) Squadron, was involved in a fatal flying accident while ferrying a P-51 Mustang, serial no. FX881, from RAF Aston Down, near Stroud, Gloucester, to Coolham ALG when the aircraft crashed just after take off.

He is buried at Northwood Cemetery, Ruislip, Middlesex, Service Plot Section H, Grave 357, and is listed on the Polish Air Force Memorial, RAF Northolt.

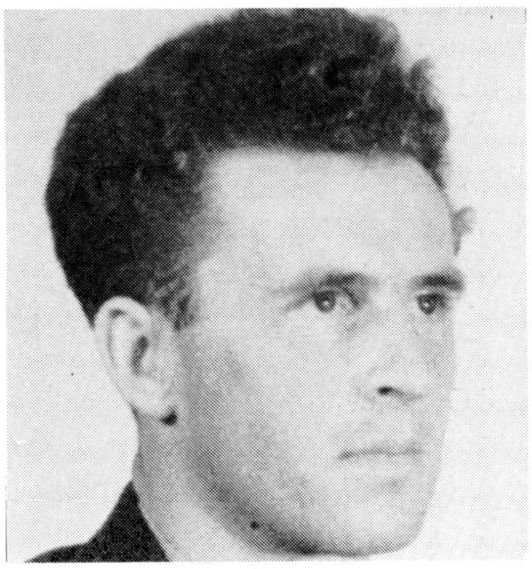
Photo: Dr. B. Nowosielski-Slepowron

Brygider, Roman Marian, P/O P-2304

Born Tarnopol, Poland 6.8.1912, died aged 31.

On 19 April 1944, P/O Brygider, a member of 315 (Polish) Squadron was involved in a fatal flying accident while evaluating a P-51 Mustang, serial no. FX896. Following a high speed dive the tail detached and the aircraft crashed close to Coolham ALG.

He is buried at Northwood Cemetery, Ruislip, Middlesex, Service Plot Section H, Grave 406, and is listed on the Polish Air Force Memorial, RAF Northolt.

Photograph unfortunately not located to date

Osieleniec, Kazimierz, F/Sgt P-780907

Born Lubin, Poland 31.7.1918, died aged 26

At 11.35hrs on 21 April 1944, F/Sgt Osieleniec, a member of 306 (Polish) Squadron, was involved in a fatal flying accident while attempting to land a P-51 Mustang, serial no. FB100 during familiarisation training at Coolham ALG. The aircraft undershot, hitting a tree stump, disabling the controls and crashed just outside the airfield.

He is buried at Northwood Cemetery, Ruislip, Middlesex, Service Plot Section H, grave 422, and is listed on the Polish Air Force Memorial, RAF Northolt.

Photo: Mrs. J. Hurndall

Nelson, Douglas Haig, P/O 54296

Born Saskatchewan, Canada 20.9.1918, died aged 26.

At about 17.30hrs on 26 April 1944, P/O Nelson, a member of 129 (Mysore) Squadron, was killed in action while flying a P-51 Mustang, serial no. FX949, on a Ranger mission. During a low level strafing attack on Beauvais airfield, the aircraft was hit by intense flak and crashed in nearby woods.

He is buried at Poix-de-la-Somme churchyard, France, Row B, Grave 32.

Photo: Mrs. M. Tyrer

Tyrer, William, LAC 1033207

Born Prescot, Lancashire, 12.3.1912, died aged 32.

At 12.50hrs on 12 May 1944, LAC Tyrer, an RAF despatch rider attached to 411 (Polish) Repair and Salvage Unit, died in a road traffic accident in West Broyle Road, Chichester, West Sussex, when his motorcycle collided with a Royal Naval lorry.

He is buried in Liverpool (West Derby) Cemetery, Section 6, Grave 565.

Photo: M. Cwynar

Calinski, Stanislaw, F/O P-2198

Born Tarnopol, Poland 9.1.1922, died aged 22.

At about 10.50hrs on 20 May 1944, F/O Calinski, a member of 315 (Polish) Squadron, was killed in action while flying a P-51 Mustang, serial no. FB179, on Ramrod 898. While dive-bombing railway marshalling yards at Charleroi, Belgium, the aircraft failed to recover from the manoeuvre.

He is buried in Creil Commonwealth Cemetery, France, Plot 2, Row 2, Grave 346, and is listed on the Polish Air Force Memorial, RAF Northolt.

Photo: R. Schuurman

Oberdak, Czeslaw, F/O P-2224

Born Krakow, Poland, 20.7.1921, died aged 23.

At about 13.00hrs on 30 May 1944 F/O Oberdak, a member of 306 (Polish) Squadron, was flying a P-51 Mustang, serial no. FX999, on Ramrod 947, acting as top cover to B-17 Fortresses, returning from operations in Germany. The aircraft developed engine trouble after encountering heavy flak and made a successful forced landing in the Zwolle area of Holland.

On 24 December he was arrested and is believed to have been executed on 8 March 1945 and is buried in a communal grave at Loenen Military Cemetery, near Apeldoorn, Holland. He is listed on the Polish Air Force Memorial, RAF Northolt.

Photo: © Crown Copyright/MOD

Geca, Boleslaw, F/Lt P-1797

Born Lubin, Poland, 23.12.1917, died aged 26.

At 10.35hrs on 7 June 1944, F/Lt Geca, a member of 306 (Polish) Squadron, was killed in action while flying a P-51 Mustang, serial no. FS169, on Ramrod 980 Part III. While dive-bombing enemy military transport in the Montfort area 19 miles south of Le Havre, the aircraft was caught in the blast from its own bombs, instantly crashing.

He is buried in Blangy Le Chateau Cemetery, France, Grave 1, and is listed on the Polish Air Force Memorial, RAF Northolt.

Photo: Dr. B. Nowosielski-Slepowron

Sworniowski, Adam, F/Lt P-1739

Born Krakow, Poland, 19.2.1918, died aged 26.

At 08.15hrs on 10 June 1944, F/Lt Sworniowski, a member of 315 (Polish) Squadron, was flying a P-51 Mustang, serial no. FX960, on Ramrod 986. While dive-bombing enemy positions, the aircraft was hit by flak. He bailed out successfully between Bayeux and Caen where it is believed he was shot by a German tank crew while attempting to escape into woods.

He is buried in Grainville, Langannierie Polish Cemetery, France, Plot 5, Row 4, Grave 9, and is listed on the Polish Air Force Memorial, RAF Northolt.

Photo: G.F. Pyle

Rigby, William Edward, W/O 177660

Born England 9.2.1923, died aged 21.

At about 15.40hrs on 14 June 1944, W/O Rigby, a member of 129 (Mysore) Squadron was killed in action while flying a P-51 Mustang, serial no. FB149, on Ramrod 998. Approximately 15 miles N/E of Caen the aircraft went missing.

He is buried at St Mymer Communal Cemetery France, Grave 2.

Photo: Mrs. S. Thomson

Green, Noel Sturt, F/Lt 122119

Born Egham, Surrey, 22.12.1922, died aged 21.

At about 11.00hrs on 17 June 1944, F/Lt Green, a member of 129 (Mysore) Squadron, was killed in action while flying a P-51 Mustang, serial no. FZ176, on Ramrod 1012, Part II. After dive-bombing a marshalling yard at Mezidon the aircraft was hit by flak and crashed.

He has no known grave and is remembered on panel 202, Runnymede Memorial, Egham, Surrey.

Photo: Mrs. C. Shiner

Payne, Stanley, F/O 158325

Born London, 11.5.1920, died aged 24.

At 18.03hrs on 17 June 1944, F/O Payne, a member of 129 (Mysore) Squadron, was killed in action while flying a P-51 Mustang, serial no. FB165, on Ramrod 1012, Part III. While taking off from Coolham ALG, fully laden, its engine lost power at low altitude, the aircraft collided with a tree, finally coming to rest in a field and was totally destroyed.

He is buried at Hills Cemetery, Horsham, Sussex, Block U, Grave 398.

Photo: M. Cwynar

Stefankiewicz, Henryk, F/Lt P-0564

Born Warsaw, Poland, 16.7.1918, died aged 26.

At about 1300hrs on 22 June 1944, F/Lt Stefankiewicz, a member of 315 (Polish) Squadron, was killed in action while flying a P-51 Mustang, serial no. FB398, on 2nd TAF Ops Order 118. During low level strafing south of Cherbourg, the aircraft was hit by intense flak and lost in the Channel.

He is remembered on the Polish Air Force Memorial at RAF Northolt.

Photo: RAF Museum Ref. No. PC 72/116/1

Hancock, Allan John, DFC & Bar, F/Lt 120710

Born Lahore, India, 26.4.1918, died aged 26.

At about 13.00hrs on 22 June 1944, F/Lt Hancock, a member of 129 (Mysore) Squadron, was killed in action while carrying out 2nd TAF Ops Order 118. During a low level strafing south of Cherbourg the aircraft was hit by intense flak and crashed.

He is buried at Bayeux War Cemetery, France, Plot 2, Row G, Grave 7.

Acknowledgements

Special thanks are extended to Beth Hamlin, Helen Shuttleworth, Rosie and Alan Woodards.

Grateful thanks are also extended to the following individuals, associations and organisations who have contributed their personal memories, photographs and assistance in compiling this book:

Mr M.M. Adam, Mr Richard Anniss, Mr R.C.B. Ashworth, Mr Tony Ayling.

Mr Richard Ball, Mr Kazimierz Barczak, Mr Jakub Bargielowski, Mr Barry Barnes, Mrs Pat Barnes, Mr John P. Bassett, S/Ldr (Ret'd) R.A. Beardsley, DFC; Mr Philip Betteridge, Mr Anthony Betts, Mr Stanislaw Blok, Mr Michael E. Booth, Mr Norman Brooks, Mr Alan Brown, Mrs Mabel Bryant, Mr Patrick Burgess, Mr Brian Bushell, Mr Brian Buss, Mr John R Byrne.

Mr. Terry Cannons, Mr W.J. Carney, Mr. John Chadwick, Mr Jim Chainey, Mr David J. Charman, Mr John Christian, Mr Ronald J. Clark, Mr David Cleverley, Mrs Joy Coleman, Mr Max A. Collett, Mr Fred Compton, Mr Richard J. Conroy, Mr Ronald M. Cook, W/Cdr (Ret'd) D.G.S.R. Cox, DFC, Bar, C de G; Mr Douglas Coxhead, AFC; Miss Hilda Craven, Mr Joe Crawshaw, Mr Charles Creasey, Mr Dave Cummins, S/Ldr (Ret'd) Michal Cwynar, DFC; Mr Stanislaw Czajkowski, Mr Bronislaw Czerwinski, Mr J. Czezowski.

Mr John Davidson, Mr Dave Davis, Mr Richard Dayton-Lewis, MChS SRch; Mr A.J. Denman, Mary Denton, Mr G.R. Dickson, Mr Stanislaw Dzialowski.

Mr Derek Edwards, Mr Mike Elliott, Mr Pierre Erkes.

Mr Rue Farnum, Mr David Francis, Mr Paul Francis, Mr W.E. Francis, Mr Joseph Foster.

Mrs Deidre Gillies, Mr Jan P. Gladysz, Mrs B. Goddard, Mr Stanislaw Gogolowski.

Mr Arthur R. Hanni, Mrs Carol Harasymowicz, Miss Patricia Hartley, Mrs Jan Hatwell, Mrs Audrey Haw, S/Ldr Charlton Haw, DFM, DFC, Order of Lenin; Mr F.T. Haw, Mrs Margaret Hazell, Mr Fred Hill, Mr Ron Hill, Mr Ray Hocking, Mrs N. Hogg, Mr Frederick Holmes, Mrs Rosemary Hook, Mr Mike Hooper, Mr Steve A. Hough, BSc, CEng, MICE; S/Ldr (Ret'd) Johnnie A. Houlton, DFC; Mr Mike Humphries, Miss Phyl Humphrey, Mrs Jean Hurndall.

Mr M.J. Ingham, FRSA.

Mr Tadeusz Jakubowski, Mr Paul Janicki, Mrs Marjorie Jarosz, S/Ldr (Ret'd) Zygmunt Jelinski, Mr Mark Jennings, Mr Len Jordan, Mr Russ Journigan.

Mr Zygmunt Kawnik, Mr Tony Kemp, Mr Jim B. Kennard, Mr J.N. King, Mr Pete King, Mr Frank Kirkham, Mr Henryk Kowalski, Mr Witold Krupa, Mr Tadeusz Krzystek.

Mrs Thelma Lapka, Mr Lech Laszkiewicz, Mr E.J. Law, Mr Brian Lee, Mr Ken C. Lee, Mr Keith Leech, Dr Lionel A. Lees, Mr A.C. Leigh, DFM, DFC; Mr Gordon Leith, Mr A.J. Lenehan, Mr Barry Lenharth, Mr Bill Lewis, Mr Lance Lidgett, Mr Richard H. Lightfine, Mr Bob Lomas, Mr Erwin Loska, Mr Leslie Lunn.

Mr Ken Markley, Mr Karol Marschall, Mr Alistair Martin-Bird, Prof J.K. Mason, CBE, MD, LLD, FRC.Path.DM; S/Ldr (Ret'd) Keith J. MacDonald, DFC; Mr Mike McCormac, Mr Paul McCue, Mr Ian McGregor, Mr Frank McPhillips, Mr Harry Matthews, Mrs V. Matthews, Mrs Eileen Millin, Mr Donald W. Mitchell, Mr Mick Monnery, Mr Tony Moor, Mr Freddy Moureau, Mr Norman Murtagh.

Mr Peter Nicholson, Mr Cedric de la Nougerede, Dr Bozydar Nowosielski-Slepowron, BSc (St A) D.I.C.PhD (St A) F.R.E.S;

Mr Desmond O'Halloran, Mr Ross H. Oliver, Mr A.F. Osborne, DFC.

Mr Ken Parfitt, Mr Dennis Parker, Mr Tom Pavey, Mr Bob Phillips, Mr George Phillips, Mr Ron Pidgley, Mrs Lindsey Pietrzak, Mr Jerzy Polak, Mr George F. Pyle.

Mr Ron Rapley, Mr. Alan Readman, Mr Robert H. Reid, Mr Neil Roggenkamp, Mr Graham Rowley, Mr Stanislaw Rudowski-Marlor.

Mrs Madge Satterly, Mr Andy Saunders, Mr Richard Schuurman, Miss Joan Scrivener, Capt Richard Sherwin, Mrs Christine L. Shiner, Mr Karol Sienkiewicz, Gen Stanislaw Skalski, DSO, DFC, VM; Mr Peter Skrzypczak, Mr Tadeusz Slon, Mr Bogdan Smidowicz, Mrs Marjorie Smith, Mr Hank Sobczynski, Mrs Joan Sologub, Mr Walter .H Standard, Mr Konrad Stembrowicz, Mr Colin Stockford, S/Ldr (Ret'd) Conrad F. Stolarkek, O.M.I; Mrs Freda Szafranska.

Mr Michael Tanner-Garbarz, Mr Geoff S. Tew, Mrs Jimmie Thomas, Mr Lucien Thomas, DFM; G/Capt (Ret'd) P.D. Thompson, DFC; Mr Alex B. Thomson, Mrs Sheila Thomson, Mr Frank Transom, Mr Ivor Turnnidge, Mrs Margaret Tyrer.

Mr Frank Webster, Mr Cliff White, Mr John R. Whitfield, Miss Karina Wiles, Mr Derek Williamson, Mr Harold Wilson, Mr Stan Wilson, Mr W.A. Winchester, Mr Adam Witko, Mr Ian J. Wood, Mr Robin Woodman, Mr Stefan Woroniecki, Mr Les Wright.

Mr Mervyn Young.

Brenzett Aeronautical Museum Trust
Commonwealth War Graves Commission
Friends of the 2nd Air Division Memorial Library, Norwich
Imperial War Museum, London
Ministry of Defence, Air Historical Branch
Ministry of Defence, RAF Personnel Management Centre
Musee Royal De L'Armee Et D'Histoire Militaire, Brussels
No. 2 Group Royal Observer Corps Association
Polish Air Force Association, Great Britain
Polish Institute and Sikorski Museum, London
Royal Air Force Association, Horsham
Royal Air Force Airfield Construction Branch Association
Royal Air Force Airfield Construction Officers Association
Royal Air Force Museum Hendon, London
Royal Air Force Regiment Association
Royal Air Force Regiment Museum
St Julians, Coolham, West SussexWest Sussex Record Office, County Hall, Chichester
West Sussex County Times
West Sussex Gazette
349 Fighter Squadron, Beauvechain, Belgium
453rd Bombardment Squadron Association
485 (New Zealand) Squadron Association

The author acknowledges that the Royal Air Force squadron badges are Crown copyright and reproduced with the permission of the Ministry of Defence.

Material used from the RAF Museum is reproduced with their permission.

The author acknowledges that the Public Record Office, Kew has custody of the following material, extracts of which were used in compiling this publication:

Air 25/195; Air 26/194; Air 27/934; Air 27/1372; Air 27/1674; Air 27/1701; Air 27/1744; Air 27/1933; Air 29/78; Air 29/109; Air 29/150; Air 29/231; Air 29/822; Air 29/826; Air 29/829; Air 29/1008; Air 37/156; Air 37/169; Air 37/171; Air 50/51; Air 50/118; Air 50/125.